What the experts are saying about Working Safe

Scott Geller knows more about the psychology of safety than anyone. In this exciting new book, he shows how to create a culture that values actively caring. It should be required reading for every safety professional, every manager, and every employee.

Terry E. McSween
President, Quality Safety Edge
Author, *The Values-Based Safety Process*

This book is a must read *for all safety professionals who desire to influence individuals and organizations for improving performance.*

Samuel J. Gualardo, CSP
Director, Corporate Safety and Health
Niagara Mohawk Power Corporation

"Actively Caring" requires us to go beyond the call of duty for our co-workers' safety and health. Working Safe *provides information and tools to increase our commitment to safety improvement, not only at work but also at home and in the community.*

Pauline Hall
Safety Specialist, Kal Kan Foods, Inc.

People are thinking more before acting; they are more involved in team goal setting; they are more empowered to ensure that safe conditions exist. This behavior-based safety process was a key factor in the shipyard receiving STAR status in the OSHA Voluntary Protection Program.

Chris Hoyer, Risk Coordinator
Elery Hudson, Senior Safety Engineer
TENNECO Newport News Shipbuilding

Working Safe *is a major and very valuable accomplishment in the cause of safety.*

H. McIlvaine Parsons, PhD
Human Resources Research Organization

Dr. Geller teaches us that safety is really in our own hands for ourselves and those around us.

Morris E. Chafetz, MD
1995 Inductee into the Health and Safety
Hall of Fame International

This book is absolutely extraordinary. It is extremely well written, accurate and factual. I highly recommend it for both the safety professional and the general business community.

John G. Cope, PhD
Professor, East Carolina University

If you've only read the rest, you haven't read the best. This book is the master tool kit for building a behavior-based safety process.

Charlie Hart
Senior Safety Coordinator, Mont Belvieu Plastics Plant
Exxon Chemical Co.

The bow of peer-reviewed psychology is bent to fire the arrow of explanation at the target of injury reduction.

Carl R. Metzgar, CSP
Chairman, The Metzgar Consulting Group

Dr. Geller has written a thoroughgoing application of psychological theory to organization systems so that safety performance can be improved and continuously refined. He courageously tells the truth—the whole truth—about human organization behavior, while remaining focused on the intent to apply behavioral science—principles, basic tools, and procedures—to build an improved safety system.

Jerry L. Burk, PhD
President, The BURK Group

This book is immediately appealing and irresistably delicious. It contains a wealth of information in highly digestible form.

Leonard Evans, D. Phil.
Author, *Traffic Safety and the Driver*

Dr. Geller breaks through the Skinner barrier—taking the work of that modern genius and making it applicable and workable for the average health-promotion and disease-prevention professional today.

William A. Smith, PhD
Executive Vice President
Academy for Educational Development

I strongly encourage anyone involved in safety management at any level to not only read this book but to study it and integrate as many of these principles as possible into your safety culture.

Robin Eidem
Manager, Professional Safety Services

Dr. Geller clearly and concisely shows executives, managers, and front line workers how to motivate themselves and others to succeed—not merely to avoid failure.

Frank J. Poliafico, RN & CEO
Emergency & Safety Programs, Inc.

An excellent book, a must for anyone interested in advancing from traditional safety to the next level. Geller has succeeded in making psychology easy to understand and apply to the safety field. It is also a very practical book with a lot of implementation ideas.

Niru Davé, CSP
Manager, Safety and Training
Champion International Corp.

WORKING SAFE

Working Safe

How to help people actively care for health and safety

by E. Scott Geller

Center for Applied Behavior Systems
Department of Psychology
Virginia Polytechnic Institute and State University

Chilton Book Company
Radnor, Pennsylvania

Published in Radnor, Pennsylvania 19089, by Chilton Book Company

Designed by Anthony Jacobson
Manufactured in the United States of America

Library of Congress Cataloging-in-Publication Data
Geller, E. Scott, 1942–
    [Psychology of safey. Selections]
    Working safe : how to help people actively care for health and
safety / E. Scott Geller.
        p.   cm.
    Selections from author's: The psychology of safety.
    Includes bibliographical references and index.
    ISBN 0-8019-8732-6
    1. Industrial safety—Psychological aspects. I. Title.
T55.3.B43G452   1996
658.3'82'01—dc20   96-7991
                                                          CIP

1 2 3 4 5 6 7 8 9 0   5 4 3 2 1 0 9 8 7 6

Past

- to my Mom, Margaret J. Scott, and Dad, Edward I. Geller, who taught me the value of learning and reinforced my need to achieve
- to B. F. Skinner and W. Edwards Deming, who developed and researched the most applicable principles in this text, and inspired me to teach them

Present

- to my wife, Carol Ann, and mother-in-law, Betty Jane, whose continuous support for 25 years made preparation to write this book possible
- to the students and associates in our university Center for Applied Behavior Systems whose data collection and analyses provided practical examples for the principles

Future

- to my daughters Krista and Karly, who I hope will someday experience the sense of accomplishment that I feel at completing this text
- to my five senior partners at Safety Performance Solutions, who I hope will continuously improve at assisting the achievement of Total Safety Cultures worldwide

Contents

Part Three: Behavior-Based Psychology

Part Four: Actively Caring for Safety

Part Five: Putting It All Together

Preface

Working Safe teaches you about the psychology of safety in order to reduce the possibility of accidental injury to yourself and to others. Psychology influences every aspect of our lives, including our safety and health. And psychology can be used to benefit almost every aspect of our lives, including our safety and health.

So what is "psychology" anyway?

My *American Heritage Dictionary*, Second College Edition, defines psychology as "1. The science of mental processes and behavior, 2. The emotional and behavioral characteristics of an individual, group, or activity" (p. 1000). Similarly, the two definitions in the *New Merriam-Webster Dictionary* are "1. The science of mind and behavior, 2. The mental and behavioral characteristics of an individual or group" (p.587).

In both dictionaries, the first definition of "psychology" uses the term "science" and refers to behavioral and mental processes. *Behaviors* are the outside, objective and observable aspects of people; *mental* or *mind* reflects our inside, subjective and unobservable characteristics. *Science* implies the application of the scientific method, or the objective and systematic analysis and interpretation of reliable observations of natural or experimental phenomena.

So what should you expect from a book on the psychology of safety? Obviously, such a book should show how psychology influences the safety and health of people. And to be useful, it should explain ways to apply psychology to improve safety and health. This is, in fact, my purpose for writing this book—to teach you how to use psychology to both explain and reduce personal injury.

As a science of mind and behavior, psychology is actually a vast field of numerous sub-disciplines. Areas covered in a standard college course in introductory psychology, for example, include: research methods, physiological foundations, sensation and perception, language and thinking, consciousness and memory, learning, motivation and emotion, human development, intelligence, personality, psychological disorders, treatment of mental disorders, social thought and behavior, environmental psychology, and industrial/organizational psychology and human factors engineering. This book does not cover all of these areas of psychology, only those directly relevant to understanding and influencing safety-related behaviors and attitudes. In addition, my coverage of information within any one sub-discipline of psychology is not comprehensive, but focuses on those aspects directly relevant to reducing injury in organizational and community settings.

This information will help you to improve safety and health in any setting, from your home to the workplace and every community location in between. You can apply the knowledge gained from reading this book in all aspects of your daily life. Most organized safety-improvement efforts occur in work environments, however, because that's where the exposure to hazardous conditions and at-risk behavior is most obvious. As a result, most (but not all) of my illustrations and examples use an industrial context. My hope is that you will see direct relevance of the principles and procedures to domains beyond the workplace.

A psychology of safety must be based on rigorous research, not common sense or intuition. This is what science is all about. Much of the psychology in self-help paperbacks, audiotapes, and motivational speeches is not founded on scientific investigation, but is presented because it sounds good and it "sells." The psychology in this book was not selected on the basis of armchair hunches but rather from the relevant research literature.

The human element of occupational health and safety has recently been an extremely popular topic at national and regional safety conferences. Safety leaders realize that reducing injuries below current levels requires increased attention to human factors. Engineering interventions and government policy have made their mark. Now it's time to work with the human dynamics of injury prevention, *the psychology of safety.*

Most attempts to deal with the human aspects of safety have been limited in scope. Many trainers and consultants claim to have answers to the human side of safety, but their solutions are too often impractical, shortsighted, or illusory. And to support their particular program, consultants, authors, and conference speakers often give unfair and inaccurate criticism of alternative methods.

Tools from behavior-based safety have been criticized in an attempt to justify a focus on people's attitudes or values. In contrast, promoters of behavior-based safety have ridiculed a focus on attitudes as being too subjective, unscientific, and unrealistic. And both behavior- and attitude-oriented approaches to injury prevention have been faulted in order to vindicate a systems or culture-based approach. The truth of the matter is that both behaviors and attitudes require attention in order to develop large-scale improvement in people's safety and health.

There are a number of books on the market that offer advice regarding the human element of occupational safety. Unfortunately, many of these texts offer a limited perspective. And I've found none comprehensive and practical enough to show how to integrate behavior- and attitude-based perspectives for a system-wide total culture transformation. This book was written to do just that, and in this regard, it is one of a kind.

Simply put, behavioral science principles provide the basic tools and procedures for building an improved safety system. But the people in a work culture need to accept and use these behavior-based techniques appropriately. This is where a broader perspective is needed, including insight regarding more subjective concepts like attitude, value, and thought processes. Recall that psychology includes the scientific study of both mind and behavior. Therefore, a practical book on the psychology of safety needs to teach science-based and feasible approaches to change what people think (attitude) and do (behavior) in order to achieve a Total Safety Culture.

I refer to a Total Safety Culture throughout this text as the ultimate vision of a safety-improvement mission. In a Total Safety Culture everyone feels responsible for safety and pursues it on a daily basis. At work, employees go beyond "the call of duty" to identify environmental hazards and at-risk behaviors. Then they intervene to correct them. And safe work practices are supported with proper recognition procedures. In a Total Safety Culture, safety is not a priority that gets shifted according to situational demands. Rather, safety is a value linked to all situational priorities.

Obviously, building a Total Safety Culture requires a long-term continuous improvement process. It involves cultivating constructive change in both the behaviors and attitudes of everyone in the culture. This book provides you with principles and procedures to make this happen. Applying what you read here might not result in a Total Safety Culture. But it's sure to make a beneficial difference in your own safety and health, and in the safety and health of others you choose to help.

I refer to helping others as *actively caring*. This book shows you how to increase the quality and quantity of your own and others' actively caring behavior. Indeed, actively caring is the key to safety improvement. The more people actively care for the safety and health of others, the less remote is the achievement of our ultimate vision—a Total Safety Culture.

WHO SHOULD READ THIS BOOK?

My editor has warned me that one book can serve only a limited audience. I know he's right, but at the same time a practical book on reducing injuries is relevant for everyone. All of us risk personal injury of some sort during the course of our days, and all of us can do something to reduce that risk to ourselves and others. Therefore, a book that teaches practical ways to do this is pertinent reading for everyone.

The average person, however, won't spend valuable time reading a book on ways to reduce personal risk for injury. Indeed, most people don't believe they are at risk for personal injury, so why should they read a book about improving safety? So while I believe everyone *should* read this book, a text on the psychology of safety is destined for a select audience: an elite group of people who are concerned about the rate of injuries in their organization or community and want to do something about it.

In order to reach more people, two versions of this book have been published: a hardcover version, *The Psychology of Safety: How to Improve Behaviors and Attitudes on the Job,* and an abbreviated softcover version: *Working Safe: How to Help People Actively Care for Health and Safety.*

The hardcover edition includes substantial research descriptions and citations to support the various principles and procedures. Thus, readers of *Working Safe* should understand that the principles and procedures explained in this edition are founded on substantial research observations as described in the hardcover edition, *The Psychology of Safety.*

The writing style and format of both versions is different than any professional text I have written or read. Most authors of professional books, including me, have been taught a particular academic or research style of writing that is not particularly enjoyable to read. When did you last pick up a nonfiction technical book for recreational or "fun" reading?

To make this book accessible to a wide variety of readers, it is written in an exciting, easy-to-read style without jargon, thanks to invaluable editorial coaching by Dave Johnson, editor of *Industrial Safety and Hygiene News.*

Each chapter includes original drawings by George Wills to illustrate concepts and add some humor to the learning process. I intersperse these drawings in my professional addresses and workshops, and audiences find them both enjoyable and enlightening. I predict that some of you will page through the book and look for these illustrations. That's a useful beginning to learning concepts and techniques for improving the human dynamics of safety. Then read the explanatory text for a second useful step toward making a difference with this information. If you then discuss the principles and procedures with others, you'll be on your way to putting this information to work in your organization, community, or home.

A TESTIMONY

Throughout this book, I include personal anecdotes to supplement the rationale of a principle or the description of a technique or process. I'd like to end this preface with one such anecdote.

In August, 1994, the Hercules Portland Plant stopped chemical production for four consecutive days so all 64 employees at the facility could receive a two-day workshop on the information presented in this book. Management had received a request for this all-employee workshop from a team of hourly workers who previously attended my two-day professional development conference sponsored by the Mt. St. Helena Section of the American Society of Safety Engineers.

Rick Moreno, a Hercules warehouse operator and hazardous materials unloader for more than 20 years, wrote the following reaction to my workshop. He read it to his coworkers at the start of the Hercules seminar. It set the stage for a most constructive and gratifying two days of education and training. If you approach the information in this book with some of the enthusiasm and optimism reflected in Rick's words, you can't help but make a difference in someone's safety and health.

Knowledge is precious. It's like trying to carry water in your cupped hands to a thirsty friend. Ideas that were crystal clear upon hearing them, tend to slip from your memory like water through the creases of your hands, and while you may have brought back enough water to wet your friend's lips, he will not enjoy the full drink that you were able to take.

And so it is with this analogy of the Total Safety Culture. Those who were there can only wet your lips with this new concept. Not a class or a program, but a safe well way to live your life that spills into other avenues of our environment.

It has no limit or boundaries as in this year, this plant. It's more like we're on our way and something wonderful is going to happen.

And even though no answers are promised or given, the avenues in which to find our own answers for our own problems will be within our reach. . . That is why it is important that everyone has the opportunity to take a full drink of the Total Safety Culture instead of having our lips wet. Something wonderful is going to happen.

This book is for you, Rick Moreno—and the many others who want to understand the psychology of safety and reduce personal injuries. Hopefully, this material will be used as a source of principles and procedures that you can return to for guidance and benchmarks along your innovative journey toward building a safer culture of more actively caring people.

E. Scott Geller
February 1996

Acknowledgments

I purchased an attractive print of a newborn colt from an artist at Galeria San Juan, Puerto Rico, in December 1992. While the artist, Jan D'Esopo, was signing my print, I asked her how long it took to complete the original.

"Twenty-five minutes or 25 years," she replied, "depending on how you look at it."

"What do you mean?" I asked.

"Well, it took me only 25 minutes to fill the canvas, but it took me 25 years of training and experience to prepare for the artistry."

I feel similarly about completing this book. While I started to write the original draft about a year ago, I've been preparing to write this book since entering the College of Wooster in Wooster, Ohio, in 1960. Almost all exams at this small liberal arts college required written discussion (rather than selecting an answer from a list of alternatives). Therefore, I received early experience and feedback at integrating concepts and research findings from a variety of sources. I was introduced to the scientific method at Wooster, and applied it to my own behavioral science research during both my junior and senior years.

Throughout five years of graduate education at Southern Illinois University in Carbondale, I developed sincere respect and appreciation for the scientific method as the key to profound knowledge. My primary areas of graduate study were learning, personality, social dynamics, and human information processing and decision making. The chairman of both my thesis and dissertation committees, Dr. Gordon F. Pitz, gave me special coaching in research methodology and data analysis, and refined my skills for professional writing.

In 1968, I was introduced to the principles and procedures of applied behavior analysis (the foundation of behavior-based safety) from one graduate course and a few visits to Anna State Hospital in Anna, Illinois, where two eminent scholars (Drs. Ted Ayllon and Nate Azrin) were conducting seminal research in this field. These learning experiences (brief in comparison with all my other education) convinced me that behavior-focused psychology could make large-scale improvements in people's lives. This insight was to have dramatic influence on my future teaching, research, and scholarship.

I started my professional career in 1969 as Assistant Professor of Psychology at Virginia Polytechnic Institute and State University (Virginia Tech). With assistance from undergraduate and graduate students, I developed a productive laboratory and research program in cognitive psychology. My tenure and promotion to Associate Professor was based entirely upon my professional scholarship in this domain. However, in the mid-1970s I became concerned that this laboratory work had limited potential for helping people. And this conflicted with my personal mission statement to make beneficial differences in people's quality of life. Therefore I turned to another line of research.

Given my conviction that behavior-based psychology has the greatest potential for solving organizational and community problems, I focused my research on finding ways to make this happen. Inspired by the first Earth Day in April 1970, my students and I developed, evaluated, and refined a number of community-based techniques for increasing environment-constructive behaviors and decreasing environment-destructive behaviors. This prolific research program culminated with the 1982 Pergamon Press publication of *Preserving the environment: New strategies for behavior change*, which I coauthored with Drs. Richard A. Winett and Peter B. Everett.

Besides targeting environmental protection, my students and I applied behavior-based psychology to a number of problem areas, including prison administration, school discipline, community theft, transportation management, and alcohol-impaired driving. In the mid-1970s we began researching strategies for increasing the use of vehicle safety belts. This led to a focus on the application of behavior-based psychology to prevent unintentional injury in organizational and community settings.

Perhaps this brief history of my professional education and research experience legitimizes my authorship of a text on the psychology of safety. However, my purpose is not so much to establish credibility as to acknowledge the vast number of individuals—teachers, researchers, colleagues, and students—who prepared me for writing this book. Critical to this preparation were our numerous research projects (since 1970), and this could not have been possible without dedicated contributions from hundreds of university students. My graduate students managed most of these field studies, and I'm truly grateful for their valuable talents and loyal efforts.

Financial support from a number of corporations and government agencies made our 25 years of intervention research possible. We received significant research funds from the Alcohol, Drug Abuse, and Mental Health Administration; the Alcoholic Beverage Medical Research Foundation; Anheuser-Busch Companies, Inc.; Centers for Disease Control and Prevention; Domino's Pizza, Inc.; Exxon Chemical Company; General Motors Research Laboratories; Hoechst-Celanese; the Motor Vehicle Manufacturers Association; Motors Insurance Corporation; the National Highway Traffic Safety Administration; the National Institute on Alcohol Abuse and Alcoholism; the National Institute for Occupational Safety and Health; the National Science Foundation; Sara Lee Knit Products; the U.S. Department of Education; the U.S. Department of Energy; the U.S. Department of Health, Education, and Welfare; the U.S. Department of Transportation; and the Virginia Departments of Agriculture and Commerce,

Litter Control, Motor Vehicles, and Welfare and Institutions. Profound knowledge is only possible through programmatic research, and these organizations made it possible for my students and me to develop and systematically evaluate ways to improve attitudes and behaviors throughout organizations and communities.

I am also indebted to the numerous guiding and motivating communications I have received from corporate and community safety professionals worldwide. Daily contacts with these individuals shaped my research and scholarship, and challenged me to improve the connection between research and application. They also provided valuable positive reinforcement to prevent burnout. It would take pages to name all of these friends and acquaintances, and then I would necessarily miss many. You know who you are—Thank you!

The advice, feedback, and friendship of two individuals—Harry Glaser and Dave Johnson—have been invaluable for my preparation to write this text. I first met Harry Glaser in September 1992, after I gave a keynote address at a professional development conference for the American Society of Safety Engineers. As Executive Vice President of Tel-A-Train, Inc., Harry decided that a video-training series on the human dynamics I presented in my talk would be useful. That was the start of an ongoing collaboration in developing videotape scripts, training manuals, and facilitator guides, all of which proved to be valuable preparation for writing this book. In particular, my relationship with Harry Glaser improved my ability to communicate the practical implications of academic research and scholarship.

Also vital to bridging the gap between research and application has been my long-term alliance and synergism with Dave Johnson, Editor of *Industrial Safety and Hygiene News*. Dave and I began learning from each other in the Spring of 1990, when I submitted my first article for his magazine. That year I submitted five articles on the psychology of safety, and Dave did substantial editing on each. Every time one of my articles was published, I learned something about communicating more effectively the bottom line of a psychological principle or procedure.

As author of more than 200 research articles and former editor of the premier research journal in the applied behavioral sciences *(Journal of Applied Behavior Analysis)*, I knew quite well how to write for a research audience in psychology. But Dave Johnson showed me that when it comes to writing for safety professionals and the general public, I had a lot to learn. And in this regard, I continue to learn from both Dave Johnson and Harry Glaser.

Dave Johnson has invited me to submit articles to *ISHN* periodically; in 1994 and 1995, I submitted an article every month. To date I've submitted 47 articles to *ISHN* and each profited immensely from Dave's suggestions and feedback. Preparing these articles laid the groundwork for this text. And Dave served as editor of this book, dedicating long hours to improving the clarity and readability of my writing. Thus, the talent and insight of Dave Johnson have been incorporated throughout this book, and I am eternally beholden to him.

Before sending my chapter drafts to Dave, I sent them to colleagues for feedback. Several of these individuals gave me useful suggestions, most notably Charlie Hart (Safety Director, Exxon Chemical Company, Mt. Belvieu, Texas), Carl Metzgar (trainer and consultant in safety and loss control, Winston-Salem, North Carolina), Mike Gilmore and Steve Roberts (Senior Partners for Safety Performance Solutions, Blacksburg, Virginia).

The drawings were created by George Wills (Blacksburg, Virginia). I think they add vitality and fun to the written presentation. I hope you agree. But without the craft and dedication of Jason Fortney, the illustrations could not have been electronically scanned for use by the publisher. In fact, Jason coordinated the final processing of this entire text, combining tables and diagrams (which he refined) with George Wills' illustrations, and the word processing from Gayle Kennedy and Candice Chevaillier.

I also sincerely appreciate the daily support and encouragement I received from my graduate students in 1995: Ted Boyce, Steve Clarke, Jason DePasquale, Kent Glindemann, and Chuck Pettinger; and from my senior partners at Safety Performance Solutions: Sherry Casali, Anne French, Mike Gilmore, Molly McClintock, and Steve Roberts. I'm sincerely grateful for the "actively caring" attention and support I received from the Chilton Book Company staff, particularly Susan Clarey for sharing the vision of this text from the start (after attending one of my two-day seminars), and Kathy Conover for patiently guiding me through the continuous improvement process. All of these people, plus many, many more, have contributed to 35 years of preparation for *Working Safe*. I thank you all very much. I'm hopeful that the synergy from all your contributions will help readers make rewarding and long-term differences in people's lives.

Part One

Orientation and Approach

Chapter 1
Choosing the Right Approach

The basic purpose of this book is outlined in Chapter 1—to explore the human dynamics of occupational health and safety, and show how they can be managed to significantly improve safety performance. The principles and practical procedures you'll learn are not based on common sense or intuition, but rather on reliable scientific investigation. Many recommendations seem counter to "pop psychology" and traditional approaches to safety. So keep an open mind while you read about the psychology of safety.

Safety professionals, team leaders, and concerned workers today scramble to find the "best" safety approach for their workplace. Typically, whatever offers the cheapest "quick fix" sells. This is not surprising, given the "lean and mean" atmosphere of the times. Programs that offer the most benefit with least effort sound best, but will they really work to improve safety over the long term?

This text will help you ask the right questions to determine whether a particular approach to safety improvement will work. More importantly, this text describes the basic ingredients needed to improve organizational and community safety. In fact, you'll find sufficient information to improve any safety process. Learning the principles and procedures described here will enable you to make a beneficial, long-term difference in the safety and health of your workplace, home, and community.

Selecting the Most Appropriate Program

With so many different approaches to safety improvement available, how can we select the best? My first thought is to ask, "What does the research indicate?" In other words, are there objective data available from program comparisons to shed light on our dilemma? Unfortunately, there are few systematic comparisons of alternative safety models. However, this doesn't stop consultants from showing us impressive results regarding the success of their approaches. Nor does it prevent them from implying (or boldly stating) that we can obtain similar fantastic results by simply following their patented "steps to success."

Keep in mind this marketing information usually comes from selected client case studies. Very few of these "success stories" were collected objectively and reliably enough to meet the rigorous standards of a professional research journal. When consultants try to sell you an approach to safety with this kind of data, ask them if they have published their results in a peer-reviewed journal. If they can show you a published report of their impressive results or a professional presentation of a program very similar to theirs, then give their approach special consideration in your selection process.

Most of the published research on safety improvement systematically evaluates whether a particular program worked in a particular situation, but it does not compare one approach with another. In other words, this research tells us whether a particular strategy is better than nothing, but offers no information regarding the relative impact of two or more different strategies to safety improvement. Such research has limited usefulness when selecting between different approaches.

An exception can be found in a 1993 review article in *Safety Science*, where Stephen Guastello (1993) summarized systematically the evaluation data from 53 different research reports of safety programs. Dr. Guastello provided rare and useful information for deciding how to improve safety. You can assume the evaluations were both reliable and valid, since each report appeared in a scientific peer-reviewed journal. All of the studies selected for his summary were conducted in a workplace setting since 1977, and each study evaluated program impact with outcome data (including number and severity of injuries).

From my reading of Dr. Guastello's article, I believe it's safe to say the behavior-based and comprehensive ergonomics approaches lead the field. Personnel selection, the most popular method (26 studies targeted a total of 19,177 employees), is among the least effective. Here are brief descriptions of the various approaches to reduce workplace injuries, ranked according to their effectiveness.

1. BEHAVIOR-BASED. Programs in this category consisted of employee training regarding particular safe and at-risk behaviors, systematic observation and recording of the targeted behaviors, and feedback to workers regarding the frequency or percentage of safe versus at-risk behavior. Some of these programs included goal setting and/or incentives to encourage the observation and feedback process.

2. COMPREHENSIVE ERGONOMICS. The ergonomics (or human factors) approach to safety refers essentially to any adjustment of working conditions or equipment in order to reduce the frequency or probability of an environmental hazard or at-risk behavior. An essential ingredient in these programs was a diagnostic survey or environmental audit by employees which led to specific recommendations for eliminating hazards that put employees at risk or promoted at-risk behaviors.

3. ENGINEERING CHANGES. This category included the introduction of robots or the comprehensive redesign of facilities to eliminate certain at-risk behaviors. It was noted, however, that the robotic interventions introduced the potential for new types of workplace injuries, like a robot catching an operator in its work envelope and impaling him or her against a structure. Thus, robotic innovations usually require additional engineering intervention such as equipment guards, emergency kill switches, radar-type sensors, and workplace re-design to prevent injury from robots.

4. GROUP PROBLEM SOLVING. For this approach, operations personnel met voluntarily to discuss safety issues and problems, and to develop action plans for safety improvement. This approach is analogous to quality circles where employees who perform similar types of work meet regularly to solve problems of product quality, productivity, and cost.

5. GOVERNMENT ACTION (IN FINLAND). Two Finnish government agencies responsible for labor production targeted the most problematic occupational groups and implemented certain action strategies. These included: a) disseminating information to work supervisors regarding the cause of workplace injuries and methods to reduce them; b) setting standards for safe machine repair and use; and c) conducting periodic work site inspections.

6. MANAGEMENT AUDITS. For the programs in this category, designated managers were trained to administer a standard International Safety Rating System (ISRS). This system evaluates workplaces based on 20 components of industrial safety. These include: leadership and administration, management training, planned inspections, task and procedures analysis, accident investigations, task observations, emergency preparedness, organizational rules, accident analysis, employee training, personal protective equipment, health control, program evaluation, engineering controls, and off-the-job safety. Managers conduct the comprehensive audit annually to develop improvement strategies for the next year. Specially certified ISRS personnel visited target sites and recognized a plant with up to five "stars" for exemplary safety performance.

7. STRESS MANAGEMENT. These programs taught employees how to cope with stressors, or sources of work stress. Exercise was often a key action strategy promoted as a way to prevent stress-related injuries in physically demanding jobs. I discuss the topic of stress as it relates to injury prevention in Chapter 6.

8. POSTER CAMPAIGNS. The two published studies in this category evaluated the accident reduction impact of posting signs that urged workers at a shipyard to avoid certain at-risk behaviors and to follow certain safe behaviors. Most signs were posted at relevant locations and gave specific behavioral instructions like "Take material for only one workday," "Gather hoses immediately after use," "Wear your safety helmet," and "Check railing and platform couplings (on scaffolds)."

For one study, safety personnel at the shipyard gave work teams weekly feedback regarding compliance with sign instructions. In the other study, environmental audits, group discussions, and structured interviews were used to develop the poster messages. Thus, it's possible factors other than the posters themselves contributed to the moderate short-term impact of this intervention approach.

9. PERSONNEL SELECTION. This popular but ineffective approach to injury prevention is based on the intuitive notion of "accident proneness." The strategy is to identify aspects of accident proneness among job applicants and then screen out people with critical levels of certain characteristics. Although this strategy sounds like a "quick fix" approach to injury prevention, this method has several problems you will readily realize as you read more in this book about the psychology of safety. Briefly, this technique has not worked reliably to prevent workplace injuries because: a) the instruments or procedures available to measure the proneness characteristics are unreliable or invalid; b) the characteristics do not carry across settings, so a person might show them at home but not at work or vice versa; and c) a person with a higher desire to take risks (such as a sensation seeker) might be more inclined to take appropriate precautions (like using personal protective equipment) to avoid potential injury.

10. "NEAR-MISS" REPORTING. This approach involved increased reporting and investigation of incidents that did not result in an injury but certainly could have under slightly different circumstances. One program in this category increased the number of corrective suggestions generated, but did not reduce injury rate. The other scientific publication in this category reported a 56-percent reduction in injury severity as a result of increased reporting of near hits[1], but the overall number of injuries did not change.

THE CRITICAL HUMAN ELEMENT

Every safety approach described above requires you to consider the human element, or the psychology of safety. Indeed, the most successful approaches, behavior-based safety and comprehensive ergonomics, directly

---

1. "Near miss" is used routinely in the workplace to refer to an incident that did not result in an injury. Since a literal translation of this term would mean the injury actually occurred, "near hit" is used throughout this text instead of "near miss."

address the human aspects of safety. The behavior-based approach targets human behavior and relies on interpersonal observation and feedback for intervention. The success of comprehensive ergonomics depends on employees observing interaction between behaviors and work situations, and then recommending feasible changes in behavior, equipment, or environmental conditions for safety achievement.

Achieving success in safety today requires concerted efforts in the realm of psychology. Safety professionals are hungry for insights. In recent years, many seminars at national and regional safety conferences purporting to teach aspects of the psychology of safety have attracted standing-room-only crowds. Just look at these titles from recent conferences of the National Safety Council or the American Society of Safety Engineers:

- "Managing Safe Behavior for Lasting Change"
- "Humanizing the Total Safety Program"
- "The Human Element in Achieving a Total Safety Culture"
- "The Psychology of Injury Prevention"
- "Behavior-Based Safety Management: Parallels with the Quality Process"
- "Behavioral Management Techniques for Continuous Improvement"
- "Improving Safety Through Innovative Behavioral and Cultural Approaches"
- "Total Safety Leadership—Empowering All Employees"
- "Managing Safety Using a Behavior-Based Approach"
- "Potholes in the Road to Behavioral Safety"
- "Achieving Total Safety Through Employee Involvement"

I attended each of these seminars and found numerous inconsistencies between presentations dealing with the same topic. Sometimes, I noted erroneous and frivolous statements, inaccurate or incomplete reference to psychological theory or research, and invalid or irresponsible comparisons between various approaches to dealing with the psychology of safety. It seemed a primary aim of several presenters was to "sell" their own particular program or consulting services by overstating the benefits of their approach and giving an incomplete or naive discussion of alternative methods or procedures.

THE FOLLY OF CHOOSING WHAT SOUNDS GOOD

The theory, research, and tools in psychology are so vast and often so complex that it can be overwhelming to decide which particular approach or strategy to use. As a result, we are easily biased by common-sense words that sound good. Valid theory, principles, and procedures founded on solid research evidence are often ignored. Today, there is an apparent endless market of self-help books, audiotapes, and videotapes addressing concepts seemingly relevant to the psychology of injury prevention. In recent years, I have listened to the following audiotapes, representing only a fraction of "pop psychology" tapes with topics relevant to the psychology of safety:

- "Coping with Difficult People" by R. M. Branson
- "Personal Excellence" by K. Blanchard
- "How to Build High Self-Esteem" by J. Canfield
- "The 7 Habits of Highly Effective People" by S. R. Covey
- "The Science of Personal Achievement" by N. Hill
- "Increasing Human Effectiveness" by R. Moawad
- "Lead the Field" by E. Nightingale
- "Unlimited Power" by A. Robbins
- "The Psychology of Achievement" by B. Tracey
- "The Psychology of Winning" by D. Waitley
- "Self-Esteem" by J. White
- "Goal Setting" by Z. Zigler

Which, if any, of these "pop psychology" audiotapes gives safety professionals the "truth"—the most effective and practical tools for dealing with the human aspects of safety? Some of the most cost-effective strategies for managing behaviors and attitudes at the personal and organizational level are not even mentioned in many of the "pop psychology" books, audiotapes, and videotapes. This might be the case not only because authors and presenters are unaware of the latest research, but also because many of the best techniques for individual and group improvement do not sound good—at least at first. The primary purpose of this text is to teach the most effective approaches for dealing with the human aspects of occupational safety and health. These principles and procedures were not selected because they sound good, but because their validity has been supported with sound research.

THE FALLACY OF RELYING ON COMMON SENSE

At a four-day "Quality Enhancement Seminar" in 1991, I heard W. Edwards Deming assert, "Experience teaches us nothing; that's why American business is in such a mess." Dr. Deming called for theory to guide objective and reliable observations, and to integrate the results from these systematic data-based experiences. Thus, while common sense is gained through biased subjective experience, scientific knowledge is gained through theory-driven objective experimentation.

Dr. Aubrey C. Daniels, a world-renowned educator and consultant in the field of organizational performance management, asserts in his 1994 book, *Bringing Out the Best in People,* that he is "on a crusade to stamp out the use of common sense in business. Contrary to popular belief there is not too little common sense in business, there is too much." (Daniels, 1994, p. 10). Dr. Daniels lists the following distinctions between common sense and scientific knowledge, reflecting the need to be cautious when relying on only common sense to deal with human aspects of occupational health and safety:

> Common-sense knowledge is acquired in ordinary business and living, while scientific knowledge must be pursued deliberately and systematically.
>
> Common-sense knowledge is individual; scientific knowledge is universal.
>
> Common-sense knowledge accepts the obvious; scientific knowledge questions the obvious.
>
> Common-sense knowledge is vague; scientific knowledge is precise.
>
> Common sense cannot be counted on to produce consistent results; applications of scientific knowledge yield the same results every time.
>
> Common sense is gained through uncontrolled experience; scientific knowledge is gained through controlled experimentation.

I've heard or read a number of psychology-related statements from motivational speakers and consultants that sound like good common sense but in fact contradict scientific knowledge. Some of these statements appear so many times in the "pop psychology" literature that they are accepted as basic truths, when in fact they cannot be substantiated with objective evidence.

How many times, for example, have you heard this statement: "Do something 21 times (or for one month) and it becomes a habit." That statement sounds reasonable, unless you know some basics about learning and the maintenance of behavior. This book will enable you to understand why such a statement about habit formation is clearly false. At this point, I hope you are open to questioning the validity of good-sounding statements that are not supported by sound research.

Research in psychology, for example, does not generally support the following common statements related to the psychology of occupational health and safety:

* Practice makes perfect.
* Spare the rod and spoil the child.
* Attitudes need to be changed before behavior will change.
* Incentives and rewards are detrimental to intrinsic motivation.
* Proper education is sufficient to motivate behavior change.
* Human nature motivates safe and healthy behavior.
* People will naturally help in a crisis.
* Rewards for not having injuries reduce injuries.

- All injuries are preventable.
- Zero injuries should be a safety goal.
- Manage only that which can be measured.

These and other common safety beliefs will be refuted in this book by scientific knowledge obtained from systematic research. Sometimes case studies will illustrate the practicality and benefits of a particular principle or procedure, but the validity of the information was not founded on case studies. The approaches presented in this text were originally discovered and verified with systematic and repeated scientific research in laboratory and field settings.

Behavior: The Scientific Approach

Many "pop psychology" self-help books, audiotapes, and motivational speeches give minimal if any attention to behavior-based approaches to personal achievement. "Behavioral control" and "behavior modification" do not sound good. The term "behavior" has negative connotations, as in "let's talk about your behavior at the party last night." Dr. B. F. Skinner, the founder of behavioral science and its many practical applications, was one of the most misunderstood and underappreciated scientists and scholars of this century, primarily because the behavior management principles he taught did not sound good. Professor Skinner and his followers have shown over and over again that behavior is motivated by its consequences, and thus behavior can be changed by controlling the events that follow behavior. But this principle of "control by consequences" doesn't sound as good as "control by positive thinking and free will." Therefore, the scientific principles and procedures from behavioral science have been underappreciated and underused.

This book teaches you how to apply behavioral science for safety achievement. The research recommends that we start with behavior. But the demonstrated validity of a behavior-based approach does not mean the better-sounding, person-based approaches should not be used. It is important to consider the feelings and attitudes of employees, since it takes people to implement the tools of behavior management. This book will teach you how certain feeling states critical for safety achievement—self-esteem, empowerment, and belongingness—can be increased by applying behavioral science. It is possible to establish interpersonal interactions and behavioral consequences in the workplace to increase important feelings and attitudes. I will show you how increasing these feeling states benefits behavior and helps to achieve safety excellence.

As illustrated in Figure 1.1, an attitude of frustration or an internal state of distress can certainly influence driving behaviors, and vice versa. Indeed, internal (unobserved) states of mind continually influence observable behaviors, while changes in observable behaviors continually affect changes in person states or attitudes. Thus, it is possible to "think a person into safe behaviors" (through education, counselling, and consensus-building exercises), and it is possible to "act a person into safe thinking" (through behavior management techniques).

Figure 1.1. Behavior influences attitude and attitude influences behavior.

In an industrial setting, it is most cost effective to target behaviors first through behavior management interventions (to be reviewed in the following chapters) implemented by employees themselves. Small changes in behavior can result in attitude change, followed by more behavior change and more desired attitude change. This spiraling of behavior feeding attitude, attitude feeding behavior, behaviors feeding attitudes and so on can lead to employees becoming totally committed to safety achievement, as reflected in their daily behavior. And all of this could start with a relatively insignificant behavior change in one employee—a "small win."

In Conclusion

In this chapter I have outlined the basic orientation and purpose of this text, which is to teach principles for understanding the human aspects of occupational health and safety, and to illustrate practical procedures for applying these principles to achieve significant improvements in organizational and community-wide safety.

The principles and procedures are not based on common sense or intuition, but rather on reliable scientific investigation. Some will contradict common folklore in "pop psychology" and require shifts in traditional approaches to the management of organizational safety. So approach this material with an open mind. Be ready to relinquish fads, fancies, and folklore for innovations based on unpopular but research-supported theory.

I promise this "psychology of safety" is based on the latest and most reliable scientific knowledge; and I promise greater safety achievement in your organization if you follow the principles and procedures presented here. Some reference to the research literature is given to verify the concepts, principles, and procedures discussed. Comprehensive reference to the research literature supporting the principles and procedures advocated in this book is available in *The Psychology of Safety* (Geller, 1996). I encourage you to read some of the original research sources to experience indirectly the rewards of scientific inquiry and distance yourself from the frivolity of common sense.

Chapter 2

Start with Theory

In this chapter we'll consider the value of theory in guiding our approaches to safety and health improvement. You'll see how a vision for a Total Safety Culture is a necessary guide to achieve safety excellence. A basic principle here is that safety performance results from the dynamic interaction of environment, behavior, and person factors. Achieving a Total Safety Culture requires attention to each of these. I make a case for integrating person-based and behavior-based psychology in order to address most effectively the human dynamics of injury prevention.

As you know, some safety efforts suffer from a "flavor of the month" syndrome. New procedures or intervention programs are tried seemingly at random, without an apparent vision, plan, or supporting set of principles. When the mission and principles are not clear, employees' acceptance and involvement suffer.

A theory or set of guiding principles makes it possible to evaluate the consistency and validity of program goals and intervention strategies. By summarizing the appropriate theory or principles into a mission statement, you have a standard for judging the value of your company's procedures, policies, and performance expectations.

It's important to develop a set of comprehensive principles on which to base safety procedures and policies. Then teach these principles to your employees so they are understood, accepted, and appreciated. This buy-in is certainly strengthened when employees or associates help select the safety principles to follow and summarize them in a company mission statement. This is theory-based safety. A critical challenge, of course, is to choose the most relevant theories or principles for your company culture and purpose, and develop an appropriate and feasible mission statement that reflects the right theory.

A Mission Statement

Several years ago I worked with employees of a major chemical company to develop the general mission statement for safety given in Figure 2.1. This

Mission Statement

A Total Safety Culture continually improves safety performance. To that end, a TSC:

- ☐ Promotes a work environment based on employee involvement, ownership, teamwork, education, training, and leadership.

- ☐ Builds self-esteem, empowerment, pride, enthusiasm, optimism, and encourages innovation.

- ☐ Reinforces the need for employees to actively care about their coworkers.

- ☐ Promotes the philosophy that safety is not a priority that can be reordered, but is a value associated with every priority.

- ☐ Recognizes group and individual achievement.

Figure 2.1. A safety achievement mission statement for the principles and procedures covered in this book.

vision for a Total Safety Culture serves as a guideline or standard for the material presented throughout this book, in the same way a corporate mission statement serves as a yardstick for gauging the development and implementation of policies and procedures.

This mission statement might not be suited for all organizations, but it is based on appropriate and comprehensive theory, supported by scientific data from research in psychology. Before developing this statement, employees learned basic psychological theories most relevant to improving occupational safety. These principles are illustrated throughout this text, along with operational (real-world) definitions.

THE VALUE OF THEORY AS "MAP"

I'd like to relate an experience to show how a theory can be seen as a map to guide us to a destination. The mission statement in Figure 2.1 reflects a destination for safety within the realm of psychology. This story also reflects the difficulty in finding the best theory among numerous possibilities.

I had the opportunity to conduct a training program at a company in Palatka, Florida. My client sent me step-by-step instructions to take me from Interstate 95 to Palatka. That was my map, limited in scope for sure, but sufficient I presumed to get the job done—to get me to the Holiday Inn in Palatka. But while at the National Car Rental desk, an attendant said my client's directions were incorrect and showed me the "correct way" with National's map of Jacksonville.

I was quick to give up my earlier theory (from my client's handwritten instructions) for this more professional display. After all, I now had a professionally printed map and directions from someone in the business of helping customers with travel plans—a consultant, so to speak. But National's map showed details for a limited area, and Palatka was not on the map. I couldn't verify the attendant's directions with the map, nor could I compare these directions with my client's very different instructions. Without a complete perspective, I chose the theory that looked best. And I got lost.

After traveling 15 miles, I began to question the "National Theory" and wondered whether my client's scribbling had been correct after all. But I stuck by my decision, and drove another ten miles before exiting the highway in search of further instruction.

I certainly needed to reach my destination that night, but motivation without appropriate direction can do more harm than good. In other words, a motivated worker cannot reach safety goals with the wrong theory or principles.

It was late Sunday night and the gas station off the exit ramp was closed, but another vehicle had also just stopped in the parking lot. I drove closer and announced to four tough-looking, grubby characters in a pick-up truck loaded with motorcycles that I was lost and wondered if they knew how to get to Palatka. None of these men had heard of Palatka, but one pulled out a detailed map of Florida and eventually found the town of Palatka. I couldn't see the details in the dark, but I accepted this new "theory" anyway, with no personal verification.

The packaging of this theory was not impressive, but my back was against the wall. I was desperate for a solution to my problem and had no other place to turn. As I left the parking lot with a new theory, I wondered whether I was now on the right track. Perhaps the theory obtained from the National Car rental attendants was correct and I had missed an exit. Whom should I believe? Fortunately, I looked beyond the slick packaging and went with the guy who had the more comprehensive perspective (the larger map). This theory got me to the Holiday Inn Palatka.

RELEVANCE TO OCCUPATIONAL SAFETY

That evening I thought about my experience and its relevance to safety. It reminded me of the dilemma facing many safety professionals when they choose approaches, programs, and consultants to help solve people problems related to safety. Theories, research, and tools in psychology are so vast and often so complex that it can be an overwhelming task to select a theory or set of principles to follow. As discussed in Chapter 1, there is a huge market of self-help books, audiotapes, and videotapes addressing concepts seemingly relevant to understanding and managing the human dynamics of safety. But many of the anecdotes, principles, and procedures given in "pop psychology" books and audiotapes are founded on limited or no scientific data. In fact, more of the material was probably used because it sounded good rather than because systematic research found it valid.

The theory that got me to Palatka was not the most professional or believable, nor was it "packaged" impressively. This doesn't mean you should avoid the slick, well-marketed approaches to occupational safety. I only wish these factors were given much less weight than scientific data.

It is relevant, though, that the more comprehensive map enabled me to find my destination. I have found that many of the human approaches to improving safety are limited in scope or theoretical foundation. Many are sold or taught as packaged programs or step-by-step procedures for any workplace culture.

In the long run, it's more useful to teach comprehensive theory and principles. On this foundation, culture-relevant procedures and interventions can be crafted by employees who will "own" and thus follow them. As the old saying goes, "Give a man a fish and you feed him for a day, teach him how to fish and you feed him for a lifetime."

At breakfast, I told the human resource manager and the safety director, the one who gave me the handwritten instructions, about my problems finding Palatka. Interestingly, each had a different theory on the best way to travel between the Jacksonville airport and Palatka. The safety director stuck with his initial instructions. The human resource manager recommended the route I eventually took. Their discussion was not enlightening; in fact it got me more confused because I didn't have a visual picture or schema (a comprehensive map) in which to fit the various approaches (or routes) they were discussing. In other words, I did not have a framework or paradigm to organize their verbal descriptions. Without a relevant theory, my experience taught me nothing, except the need for an appropriate theory—in this case a map.

A theory should serve as the map that provides direction to meet a specific safety challenge. Obviously it's important to teach the basic theory to everyone who must meet the challenge. Then it's a good idea to have an employee task force summarize the theory in a safety mission statement. When the workforce understands the theory and accepts the summary mission statement, intervention processes based on the theory will not be viewed as "flavor of the month," but as an action plan to bring the theory to life.

When employees appreciate and affirm the theory, they will get involved in designing and implementing the action steps. They will also suggest ways to refine or expand action plans and theory on the basis of systematic observations or scientific evidence. This is the best kind of continuous improvement.

A BASIC MISSION AND THEORY

The mission statement in Figure 2.1 reflects the ultimate in safety—a Total Safety Culture, in which:

- Everyone feels responsible for safety and does something about it on a daily basis.
- People go beyond the call of duty to identify unsafe conditions and at-risk behaviors, and they intervene to correct them.
- Safe work practices are supported intermittently with rewarding feedback from both peers and managers.
- People "actively care" continuously for the safety of themselves and others.
- Safety is not considered a priority that can be conveniently shifted depending on the demands of the situation; rather safety is considered a value linked with every priority of a given situation.

This Total Safety Culture mission is much easier said than done, but it is achievable through a variety of safety processes rooted in the disciplines of engineering and psychology. Generally, a Total Safety Culture requires continual attention to three domains:

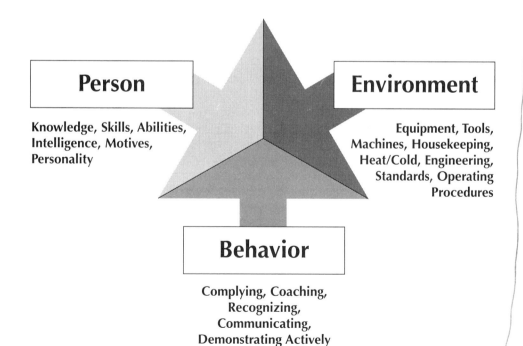

Figure 2.2. A Total Safety Culture requires continual attention to three types of factors.

1. Environment factors (including equipment, tools, physical layout, procedures, standards, and temperature);

2. Person factors (including people's attitudes, beliefs, and personalities); and

3. Behavior factors (including safe and at-risk work practices, as well as going beyond the call of duty to intervene on behalf of another person's safety).

These factors are dynamic and interactive. Changes in one domain eventually impact the other two. For example, behaviors that reduce the probability of injury often involve environmental change, and lead to attitudes consistent with the safe behaviors. The behavior and person factors represent the human dynamics of occupational safety and are addressed in this book. The basic principle here is that behavior-based and person-based factors need to be addressed in order to achieve a Total Safety Culture. These two divergent approaches to understanding and managing the human element represent the psychology of injury prevention.

Behavior-Based versus Person-Based Approaches

Paying attention to only behavior-based factors (the observable activities of people) or to only person-based factors (unobservable feeling states or attitudes of people) is like using a limited map to find a destination, as with my attempt to find Palatka, Florida. The mission to achieve a Total Safety Culture requires a comprehensive framework—a complete map of the relevant psychological territory. Figure 2.3 illustrates the complex interaction of environment, person, and behavior factors.

There are myriad opinions and recommendations on how the psychology of safety can be used to produce beneficial changes in people and organizations. Most can be classified into one of two basic approaches: person-based and behavior-based. In fact, most of the numerous psychotherapies available to treat developmental disabilities and psychological disor-

THIS EXTREMELY DANGEROUS BRAIN SURGERY IS AIDED BY THE MOST HIGH-TECH, STATE-OF-THE-ART PROBING DEVICE EVER ENGINEERED... BUT YOU KNOW... THE SURGEON TOTALED HIS PORSCHE LAST NIGHT.

Figure 2.3. Performance results from the dynamic interaction of environment, behavior, and person factors.

ders, from neurosis to psychosis, can be classified as essentially person-based or behavior-based.

That is, most psychotherapies focus on changing people either from the inside ("thinking people into acting differently") or from the outside ("acting people into thinking differently"). Person-based approaches attack individual attitudes or thinking processes directly. They teach clients new thinking strategies or give them insight into the origin of their abnormal or unhealthy thoughts, attitudes, or feelings. In contrast, behavior-based approaches attack the clients' behaviors directly. They change relationships between behaviors and their consequences. This text will show you how to integrate relevant principles from these two psychological approaches in order to achieve a Total Safety Culture.

THE PERSON-BASED APPROACH

Imagine you see two employees pushing each other in a parking lot, as a crowd gathers around to watch. Is this aggressive behavior, horseplay, or mutual instruction for self defense? Are the employees physically attacking each other to inflict harm, or does this physical contact indicate a special friendship and mutual understanding of the line between aggression and play? Perhaps if you watch longer, and pay attention to verbal behavior, you'll decide whether this is aggression, horseplay, or a teaching/learning demonstration.

However, a truly accurate account might require you to assess each individual's personal feelings, attitudes, or intentions. It's possible, in fact, that one person was being hostile while the other was just having fun, or the contact started as horseplay and progressed to aggression.

This scenario illustrates a basic premise of the person-based approach. Focusing only on observable behavior does not explain enough. People are much more than their behaviors. Concepts like intention, creativity, intrinsic motivation, subjective interpretation, self-esteem, and mental attitude are essential to understanding and appreciating the human dynamics of a problem. Thus, a person-based approach in the workplace applies surveys, personal interviews and focus-group discussions to find out how individuals feel about certain situations, conditions, behaviors, or personal interactions.

Humanism is the most popular person-based approach today, as evidenced by the current market of "pop psychology" videotapes, audiotapes, and self-help books. Some current popular industrial psychology tools—such as the Myers-Briggs Type Indicator and other trait measures of personality, motivation, or risk-taking propensity—stem from psychoanalytic theory and practice. The key principles of humanism found in most "pop psychology" approaches to increasing personal achievement are:

1. Everyone is unique in numerous ways. The special characteristics of individuals cannot be understood or appreciated by applying general principles or concepts, such as the behavior-based principles of performance management, or the permanent personality trait perspective of psychoanalysis.

2. Individuals have far more potential to achieve than they typically realize and should not feel hampered by past experiences or present liabilities.

3. The present state of an individual in terms of feeling, thinking, and believing is a critical determinant of personal success.

4. One's self-concept influences mental and physical health, as well as personal effectiveness and achievement.

5. Ineffectiveness and abnormal behavior result from large discrepancies between one's real self ("who I am") and ideal self ("who I would like to be").

6. Individual motives vary widely and come from within a person.

Readers familiar with the writings of W. Edwards Deming (1986, 1993) and Stephen R. Covey (1989, 1990) will recognize these eminent industrial consultants as humanists, or advocates of a person-based approach.

THE BEHAVIOR-BASED APPROACH

The behavior-based approach to applied psychology is founded on behavioral science as conceptualized and researched by B. F. Skinner (1938, 1974). In his experimental analysis of behavior, Skinner rejected for scientific study unobservable factors such as self-esteem, intentions, and attitudes. He researched only observable behavior and its social, environmental, and physiological determinants. The behavior-based approach starts by identifying observable behaviors targeted for change and the environmental conditions that can be manipulated to influence the target behavior(s) in desired directions.

Behaviorism has effectively solved environmental, safety, and health problems in organizations and throughout entire communities. You first define the problem in terms of relevant observable behavior, and then design and implement an intervention process to decrease behaviors causing the problem and/or increase behaviors that can alleviate the problem.

CONSIDERING COST-EFFECTIVENESS

When people act in certain ways they usually adjust their mental attitude and self-talk to parallel their actions; and when people change their attitudes, values, or thinking strategies, certain behaviors change as a result. Thus, person-based and behavior-based approaches to changing people can influence both attitudes and behaviors, either directly or indirectly. Most parents, teachers, first-line supervisors, and safety managers use both approaches in their attempts to change a person's knowledge, skills, attitudes, or behaviors:

> When we lecture, counsel, or educate others in a one-on-one or group situation, we are essentially using a person-based approach
>
> When we recognize, correct or discipline others for what they have done, we are operating from a behavior-based perspective

Unfortunately, we're not always effective with our person-based or behavior-based change techniques, and often we don't know whether our intervention worked as intended. In order to apply person-based techniques to psychotherapy, clinical psychologists receive specialized therapy or counseling training for four years or more, followed by an internship of at least a year. This intensive training is needed because tapping into an individual's perceptions, attitudes, and thinking styles is a demanding and complex process. Also, internal dimensions of people are extremely difficult to measure reliably, making it cumbersome to assess therapeutic progress and obtain straightforward feedback regarding therapy skills. Consequently, the person-based therapy process can be very time-consuming, requiring numerous one-on-one sessions between professional therapist and client.

In contrast, behavior-based psychotherapy was designed to be administered by individuals with minimal professional training. From the start, the idea was to reach people where problems occur—in the home, school, rehabilitation institute, and workplace, for example—and teach parents, teachers, supervisors, friends, or co-workers the behavior-change techniques most likely to work under the circumstances.

More than three decades of research have shown convincingly that this on-site approach is cost effective, primarily because behavior-change techniques are straightforward and relatively easy to administer, and because intervention progress can be readily monitored by the ongoing observation of target behaviors. By obtaining objective feedback on the impact of intervention techniques, a behavior-based process can be continually refined or altered.

Integrating Approaches

A common perspective, even among psychologists, is that humanists and behaviorists are complete opposites. Behaviorists are considered cold, objective, and mechanistic, operating with minimal concern for people's feelings. In contrast, humanists are thought of as warm, subjective and caring, with limited concern for directly changing another person's behavior or attitude. The basic humanistic approach is termed "nondirective" or "client-centered."

Therapists, counselors, or coaches do not directly change their clients, but rather provide empathy and a caring and supportive environment for enabling clients to change themselves, from the inside out.

Given the foundations of humanism and behaviorism, it is easy to build barriers between person-based and behavior-based perspectives and assume you must follow one or the other when designing an intervention process. In fact, many consultants in the safety management field market themselves as using one or the other approach, but not both. It is my firm belief that these approaches need to be integrated in order to truly understand the psychology of safety and build a Total Safety Culture.

In Conclusion

Theory or basic principles are needed to organize research findings and guide our approaches to improve the safety and health of an organization. Similarly, a vision for a Total Safety Culture incorporated into a mission statement is needed to guide us in developing action plans to achieve safety excellence.

When employees understand and accept the mission statement and guiding principles, they become more involved in the mission. The action plan will not be viewed as one more "flavor of the month," but as relevant to the right principles and useful for achieving shared goals. Indeed, the workforce will help design and implement the action plans. This is crucial for developing a Total Safety Culture.

A basic principle introduced in this chapter is that the safety performance of an organization results from the dynamic interaction of environment, behavior, and person factors. The behavior and person dimensions represent the human aspect of industrial safety, and reflect two divergent approaches to understanding the psychology of injury prevention. The behavior-based approach is more cost-effective than the person-based approach in effecting large-scale change. But it cannot be effective unless the work culture believes in the behavior-based principles and willingly applies them to achieve the mutual safety mission. This involves person-based psychology, which means a Total Safety Culture requires us to integrate person-based and behavior-based psychology.

Chapter 3

Paradigm Shifts for Total Safety

This chapter outlines ten new perspectives we need to adopt in order to exceed current levels of safety excellence and reach our ultimate goal—a Total Safety Culture. The traditional three Es of safety management—engineering, education, and enforcement—have only gotten us so far. A Total Safety Culture requires understanding and applying three additional Es—empowerment, ergonomics, and evaluation.

Safety in industry has improved dramatically in this century. Let's examine the evolution of accident prevention to see how this was accomplished. The first systematic research began in the early 1900s, and focused on finding the psychological causes of accidents. It assumed people were responsible for most accidents and injuries, usually through mental errors caused by anxiety, attitude, fear, stress, personality, or emotional state. Reducing accidents was typically attempted by "readjusting" attitude or personality, usually through supervisor counseling or discipline.

This so-called "psychological approach" held that certain individuals were "accident prone." By removing these workers from risky jobs or by disciplining them to correct their attitude or personality problems, it was thought that accidents could be reduced. But as I discussed in Chapter 1, this focus on accident-proneness has not been effective, partly because reliable and valid measurement procedures are not available. Also, the person factors contributing to accident proneness are probably not consistent characteristics or traits in people, but vary from time to time and situation to situation.

The "Old" Three Es

Enthusiasm for the early "psychological approach" waned because of the difficulty of measuring its impact. In addition, the seminal research and scholarship of William Haddon (1968) suggested that engineering changes held the most promise for large-scale, long-term reductions in injury severity.

As the first administrator of the National Highway Safety Bureau (now the National Highway Traffic Safety Administration), Dr. Haddon was able to turn his theory and research into the first federal automobile safety standards. Haddon believed injury is caused by delivering excess energy to the body, and that injury prevention depends on controlling that energy. The prevention focus now shifted to engineering and epidemiology, and resulted in developing personal protective equipment (PPE) for work and recreational environments, as well as standards and policy regarding the use of PPE. In vehicles, Haddon's basic theory eventually led to collapsible steering wheels, padded dashboards, head restraints, and airbags in automobiles.

This brief history of the safety movement in the U.S. explains why engineering is the dominant paradigm in industrial health and safety, with secondary emphasis on two additional Es—education and enforcement. Over the past several decades, the basic protocol for reducing injury has been to:

1. Design the safest equipment, environmental settings, or protective devices

2. Educate people regarding the use of the engineering interventions

3. Use discipline to enforce compliance with recommended safe work practices

Thanks to this paradigm, most safety professionals are safety *engineers*, who commonly advocate that "Safety is a condition of employment."

The three Es have dramatically reduced injury severity in the workplace, at home, and on the road. Let's take a look at motor vehicle safety for a minute. The Government Accounting Office has estimated conservatively that the early automobile safety standards ushered through Congress by Dr. Haddon had saved at least 28,000 American lives by 1974. In addition, the state laws passed in the 1980s requiring use of vehicle safety belts and child safety seats have saved countless more lives. Many more lives would be saved and injuries avoided if more people buckled up and used child safety seats for their children.

The current rate of safety belt use in the U.S. is about 67 percent, a dramatic improvement from the 15 percent prior to statewide interventions, including belt-use laws, campaigns to educate people about the value of safety-belt use, and large-scale enforcement blitzes by local and state police officers.

There is still much room for improvement, especially considering that most of the riskiest drivers still don't buckle up. Each year since 1990, the U.S. Department of Transportation has set nationwide belt-use goals of 70 percent, but to date this goal has not been met—at least over the long term.

It seems the effectiveness of current methods to increase the use of this particular type of PPE has plateaued or asymptoted below 70 percent.

Now let's turn our attention to industry. I've worked with many corporate safety professionals over the years who say their plants' safety performance has reached a plateau. Yes, their overall safety record is vastly better than it once was. But continuous improvement is elusive. A frantic search for ways to take safety to the next level has not paid off. The old "three Es" will not get us there.

A certain percentage of people keep falling through the cracks. Keep on doing what you're doing and you'll keep on getting what you're getting. As I heard Dr. Deming (1991) say many times, "Goals without method, what could be worse?"

The New Three Es

I'm going to introduce you to three new Es in this book. Ergonomics. Empowerment. Evaluation. I certainly don't suggest abandoning tradition. We need to maintain a focus on engineering, education, and enforcement strategies. But to get beyond current plateaus and reach new heights in safety excellence, we must attend more competently to the psychology of injury prevention. These three new Es suggest specific directions or principles.

ERGONOMICS

As discussed in Chapter 1, ergonomics requires careful study of relationships between environment and behavior, as well as developing action plans (such as equipment work orders, safer operating procedures, training exercises) to avoid possible acute or chronic injury from the environment/behavior interaction. This requires consistent and voluntary participation by those who perform the behaviors in the various work environments. These are usually line operators or hourly workers in an organization, and their participation will happen when these individuals are empowered.

Throughout this book I discuss ways to develop an empowered work culture, and I explain procedures for involving employees in ergonomic interventions.

EMPOWERMENT

Some operational definitions of the traditional three Es for safety (especially enforcement) have been detrimental to employee empowerment. Many supervisors have translated "enforcement" into a strict punishment approach, and the result has turned off many employees to safety programs. These workers may do what is required, but no more. Some individuals who feel especially controlled by safety regulations might try and beat the system, and success will likely bring a sense of gratification or freedom (as illustrated in Figure 3.1.)

EVALUATION

The third new E essential to achieving a Total Safety Culture is evaluation. Without appropriate feedback or evaluation, practice does not make perfect. Thus, we need the right kind of evaluation process. Later in this book, especially in Chapter 15, I detail proce-

Figure 3.1. Some top-down rules have undesirable side effects.

dures for conducting the right kind of comprehensive evaluation. Right now, what's important to understand is that some traditional methods of evaluation actually decrease or stifle empowerment. This calls for changing some safety measurement paradigms.

Remember the need for a guiding theory or set of principles? Basic theory from person-based and behavior-based psychology suggests a shift to new safety paradigms. And, these paradigm shifts provide a new set of guiding principles for achieving new heights in safety excellence.

I want to define ten basic changes in belief, attitude, or perception that are needed to develop the ultimate Total Safety Culture. These shifts require new principles, approaches, or procedures, and will result in different behaviors and attitudes among top managers and hourly workers. Empowerment will increase throughout the work culture.

1. FROM OSHA REGULATIONS TO CORPORATE RESPONSIBILITY. Many safety activities and programs in U.S. industry are driven by OSHA (the U.S. Occupational Safety and Health Administration) or MSHA (the Mine Safety and Health Administration) rather than by the employers and employees who can benefit from a safety process. In other words, many in industry do "safety stuff" because the government requires it—not because it was their idea and initiative.

But people are more motivated and willing to go beyond the call of duty when they are achieving their own self-initiated goals. Ownership, commitment, and proactive behaviors are less likely when you're working to avoid missing goals or deadlines set by someone else. This statement is intuitive.

Just compare your own motivation when working for personal gain versus someone else's gain, or when working to earn a reward versus to avoid a penalty.

The language used to define safety programs and activities influences personal participation. Remember, we can act ourselves into an attitude. So it makes sense to talk about safety as a company mission that is owned and achieved by the very people it benefits. A safety process is not intended to benefit federal regulators. Let's work to achieve a Total Safety Culture for the right reasons.

2. FROM FAILURE ORIENTED TO ACHIEVEMENT ORIENTED. If you strive to meet someone else's goals rather than your own, you'll probably develop an attitude of "working to avoid failure" rather than "working to achieve success." But we are more motivated by achieving success than avoiding failure. If you have a choice between earning positive reinforcers (rewards) or avoiding punishers (penalties), you'll probably choose the positive reinforcement situation. Plus, if you feel controlled by punishment or negative reinforcement, you often procrastinate and take a reactive rather than a proactive stance.

This principle helps explain why more continuous and proactive attention goes to productivity and quality than to safety. Productivity and quality goals are typically stated in achievement terms, and gains are tracked and recorded as individual or team accomplishments, sometimes followed by rewards or recognition awards.

In contrast, safety goals are most often stated in negative reinforcement terms. How many times have you heard: "We will reach our safety goal after another month without a lost-time injury." And "keeping score" in safety means tracking and recording losses or injuries.

Measuring safety with only records of injuries not only limits evaluation to a reactive stance, it also sets up a negative motivational system that is apt to take a back seat to the positive system used for productivity and quality. Giving safety an achievement perspective (like production and quality) requires a different scoring system, as indicated by the next paradigm shift.

3. FROM OUTCOME FOCUSED TO BEHAVIOR FOCUSED. Companies are frequently ranked according to their OSHA recordables and lost-time injuries. Within companies, work groups or individual workers earn safety awards according to outcomes—those with the lowest numbers win. Offering incentives for fewer injuries, for instance, can often reduce the reported numbers while not improving safety. Pressure to reduce outcomes without changing the process (or ongoing behaviors) often causes employees to cover up their injuries.

How many times have you heard of an injured employee being driven to work each day to sign in and then promptly returned to the hospital or home to recuperate? This keeps the outcome numbers low but does more harm than good to the corporate culture. Likewise, failure to report even a minor first-aid case prohibits key personnel from correcting the factors that led to the incident. A scoring system based on what people do for safety (as in a behavior-based process) not only attacks a root cause of most work injuries, it can also be achievement oriented. This puts safety in the same motivational framework as productivity and quality.

Figure 3.2. Safety reward programs should pass the "dead man's" test.

A misguided emphasis on outcomes rather than process is illustrated in Figure 3.2. Although the idea of a dead person receiving a safety award is clearly ridiculous, this type of incentive/reward process is quite common in American industry. Such programs often bring down numbers by influencing the *reporting* of injuries, but rarely do they benefit the safety processes which control results.

In Chapter 10, I explain principles for establishing an incentive/reward process to motivate the kinds of safety processes that influence outcomes. For now, just recognize and appreciate the advantage of focusing on achieving process improvements over working to avoid failure. This is especially true if a failure-oriented goal is remote, such as a plant-wide reduction in injuries, and thus might be perceived as uncontrollable.

Safety can be on equal footing with productivity and quality if it is recorded and tracked with an achievement score perceived by employees as directly controllable and obtainable. This occurs with a focus on the safety processes that can decrease an organization's injury rate, as well as an ongoing measurement system that continuously tracks safety accomplishments and displays them to the workforce. This book shows you how to do this.

4. FROM TOP-DOWN CONTROL TO BOTTOM-UP INVOLVEMENT. As I discussed when introducing three new *E*s, a Total Safety Culture requires continual involvement from operations personnel such as hourly workers. After all, these are the people who know where safety hazards are located and when the at-risk behaviors occur. Also, they can have the most influence in supporting safe behaviors and correcting at-risk behaviors and conditions. In fact, the ongoing processes involved in developing a Total Safety Culture need to be supported from the top but driven from the bottom. This is more than employee participation; it is employee ownership, commitment, and empowerment.

5. FROM RUGGED INDIVIDUAL-ISM TO TEAMWORK. An employee-driven safety process requires teamwork founded on interpersonal trust, synergy, and win/win contingencies. However, from childhood most of us have been taught an individualistic, win/lose perspective, supported by such popular slogans as "You have to blow your own horn," "Nice guys finish last," "No one can fill your shoes like you," and "It's the squeaky wheel that gets the grease." Grades in school, the legal system, and many sports also ori-

PSST! M'AM! - SEEN A LUMP OF CHEESE AROUND HERE? I'M BEING TIMED.

Figure 3.3. Some work environments create barriers to synergy.

ent us to think individualism and win/lose. This is why a true team approach to safety does not come easily.

Figure 3.3 illustrates a competitive situation quite common in the workplace. Although some office environments were originally designed to promote more open communication and group interaction, physical and psychological barriers have often been erected to maintain privacy and an individualistic atmosphere. This partially results from work systems that offer more rewards for individual than group achievement. Processes and systems can be implemented to promote group behaviors and interdependence over individual behaviors and independence. These processes and contingencies are emphasized throughout this book, since a Total Safety Culture requires more teamwork than rugged individualism.

6. FROM A PIECEMEAL TO A SYSTEMS APPROACH. The long-term improvements of a Total Safety Culture can only be achieved with a systems approach, including balanced attention to all aspects of the corporate culture. Dr. Deming (1986, 1993) emphasized that total quality can only be achieved through a systems approach, and of course the same is true for safety. As I discussed earlier in Chapter 2, three basic domains need attention when designing and evaluating safety processes and when investigating the root causes of near hits and injuries:

1. Environment factors such as equipment, tools, machines, housekeeping, heat/cold, engineering;

2. Person factors such as employees' knowledge, skills, abilities, intelligence, motives, personality; and

3. Behavior factors such as complying, coaching, recognizing, communicating, and "actively caring."

Two of these system variables involve human factors. Each generally receives less attention than the environment, mostly because it's more difficult to visibly measure the outcomes of efforts to change the human factors. Some human factors programs focus on behaviors (as in behavior-based safety); others focus on attitudes (as in a person-based approach). A Total Safety Culture integrates these two approaches.

7. FROM FAULT FINDING TO FACT FINDING. Blaming an individual or group of individuals for an injury-producing incident is not consistent with a systems approach to safety. Instead, an injury or near hit provides an opportunity to investigate facts from all aspects of the system that could have contributed to the incident. Immediate environment, person, and behavior factors should be explored for root causes, and numerous historical factors should also be considered. For example:

How common was the at-risk behavior?

How many individuals observed the at-risk behavior without intervening?

And what aspects of operations and the management system supported the at-risk behavior?

Several years ago I helped a company investigate an injury which occurred when an employee slipped on a metal plate covering a large hole on scaffolding three stories above a concrete floor. The worker fell through the hole, and would have fallen three stories were he not able to throw out his arms and catch himself on the sides of the hole. He suffered painful cuts and scrapes, but obviously the injury could have been much worse.

Whose fault was it? What was the root cause of this incident? It was tempting to finger the welder who failed to secure the metal plate over the hole as the culprit—the root cause. This would be piecemeal and fault finding. Instead a fact finding, systems approach was followed. What were the management demands, for example, that led to careless work by the welder? How many individuals had stepped on the plate, noticed that it was loose, and did not report their near hit? (The investigation revealed that numerous employees had been aware of the loose plate.) What environment or person factors prevented people from reporting their near hit with the loose plate? What processes should be put in place to facilitate observing, reporting, and correcting environmental hazards like the one contributing to this injury?

This is obviously only a partial list of questions related to a systems-level, fact-finding investigation. But I hope it's clear that answering these questions will be far more constructive than finding an individual to blame and perhaps punishing him or her for the mishap.

In this case, I doubt there was one root cause. Indeed, efforts to search for the root cause of an injury can be fruitless and lead to more fault finding. Many environment, behavior, and person factors contribute to the system that causes workplace injuries. Piecemeal corrective action is narrow-minded and short-sighted; systems-level corrective action can have a large-scale, long-term impact.

8. FROM REACTIVE TO PROACTIVE. Investigating events preceding an incident, be it a near hit or an injury, demonstrates the need to think and act proactively. Unfortunately, a proactive stance is extremely difficult to maintain, especially in a corporate culture that is increasingly complex and demanding. There is a higher and higher price tag on "free time." With barely enough time to react sufficiently to crises each day, how can we find time to be proactive?

9. FROM QUICK FIX TO CONTINUOUS IMPROVEMENT. There is no quick-fix answer to this challenging question. "Proactive" can be substituted for "reactive" only with a systems perspective and an optimistic attitude of continuous improvement through increased employee involvement. Understanding the psychology of safety can be a great aid here. The principles and procedures described in this book will enable you to influence incremental changes in work practices and attitudes that can prevent an injury. This represents a proactive, continuous improvement paradigm, which will surely improve your safety performance.

10. FROM PRIORITY TO VALUE. "Safety is a priority." This is probably the most common safety slogan found in workplaces and voiced by safety leaders. I've seen signs, pens, buttons, hats, T-shirts, and note pads with this message. No wonder safety and health professionals are surprised when I say that safety should not be a priority. To justify my case, I offer the following explanation:

Think about a typical workday morning. We all follow a prioritized agenda, often a standard routine, before traveling to work. Some people eat a hearty breakfast, read the morning newspaper, take a shower, and wash dishes. Others wake up early enough to go for a morning jog before work. Some grab a roll and a cup of coffee, and leave their home in disarray until they get back in the evening.

In each of these scenarios the agenda—the priorities—are different. Yet there is one common activity. It's not a priority but a basic value. Do you know what it is? One morning you wake up late. Perhaps your alarm-clock failed. You have only 15 minutes to prepare for work. Your morning routine changes drastically. Priorities must be rearranged. You might skip breakfast, a shower, or a shave. Yet every morning schedule still has one item in common. It's not a priority, capable of being dropped from a routine due to time constraints or a new agenda. No, this particular morning activity represents a value which we've been taught as infants, and it's never compromised. Have you guessed it by now? Yes, this common link in everyone's morning routine, regardless of time constraints, is "getting dressed."

This simple scenario shows how circumstances can alter behavior and priorities. Actually, labeling a behavior a "priority" implies that its order in a hierarchy of daily activities can be rearranged. How often does this happen at work? Does safety sometimes take a "back seat" when the emphasis is on other priorities such as production quantity or quality?

Enduring Values

It's human nature to shift priorities, or behavioral hierarchies, according to situational demands or contingencies. But values remain constant. The early morning anecdote illustrates that the activity of "getting dressed" is a value that is never dropped from the routine. Shouldn't "working safely" hold the same status as "getting dressed"? Safe work practices should occur regardless of the demands of a particular day. Safety should be a value linked with every activity or priority in a work routine. Safe work should be the enduring norm, whether the current focus is on quantity, quality, or cost-effectiveness as the "number one priority."

The ultimate aim of a Total Safety Culture is to make safety an integral aspect of all performance, regardless of the task. Safety should be more than the behaviors of "using personal protective equipment," more than "locking out power" and "checking equipment for potential hazards," and more than "practicing good housekeeping." Safety should be an unwritten rule, a social norm, that workers follow regardless of the situation. It should become a value that is never questioned—never compromised.

In Conclusion

This chapter describes ten shifts in perspective needed to go beyond current levels of safety excellence. The first nine could be considered goals for achieving a Total Safety Culture. The tenth—making safety a value—is not something that can be readily measured and tracked. It's the ideal vision for our safety mission.

Here's how the new paradigms fit together: Your safety achievement process should be considered a company responsibility, not a regulatory obligation. It should be achievement oriented with a focus on behavior, supported by all managers and supervisors but driven by the line workers or operators through teamwork. A systems approach is needed, which leads to a fact-finding perspective, a proactive stance, and a commitment to continuous improvement.

These new perspectives reflect new principles to follow, new procedures to develop and implement. This "new safety work" will lead to different perceptions, attitudes, and even values. Ultimately, the tenth paradigm shift can be reached. When safety goes from priority to value, it won't be compromised at work, at home, or on the road. Naturally, numerous injuries will be prevented and lives saved everyday. This vision should motivate each of us to be active in the safety achievement process. This book helps you define your role.

Part Two

Human Barriers to Safety

Chapter 4

The Complexity of Human Nature

Safety is usually a continuous fight with human nature. This chapter explains why. Understanding this basic point will lead to less victim-blaming and fault-finding when investigating an injury. Instead, we'll be able to find factors in the system that can be changed in advance to prevent injuries at work, at home, and throughout the community.

"All injuries are preventable."

"It's human nature to work safely."

"Safety is just common sense."

"Safety is a condition of employment."

Read these familiar statements and you get the idea that working safely is easy or natural. Nothing could be further from the truth.

In fact, it's often more convenient, more comfortable, more expedient, and more common to take risks than to work safely. And past experience usually supports our decisions to choose the at-risk behavior, whether we're working, traveling, or playing. So we're often engaged in a continuous fight with human nature to motivate ourselves and others to avoid those risky behaviors and maintain safe ones.

Let's consider what holds us back from choosing the safe way, whether it's following safe operating procedures, driving our automobile, or using personal protective equipment.

When I ask safety professionals, corporate executives, or hourly workers what causes work-related injuries, I get long and varied lists of factors. Actually, each list is quite similar. After all, everyone experiences events, attitudes, demands, distractions, responsibilities, and circumstances that get in the way of performing a task safely.

have been in situations where we weren't sure how to perform

ve lacked training. Maybe the surrounding environment was not
as ~~be~~   could be.

Demands from a supervisor, co-worker, or friend put pressure on us to take a short cut or risk. Maybe it was inconvenient or uncomfortable to follow all of the safety procedures.

It's possible a physical condition (fatigue, boredom, drug impairment) influenced at-risk performance. There are other factors. Have you ever been unsafely distracted by external stimuli, like another person's presence? Or by an internal state, like personal thoughts or emotions?

Can you remember a time when you just didn't feel like taking the extra time to be safe? And I'm sure you've experienced the "macho" attitude, from yourself or others, that "It will never happen to me." Fortunately, it's rare that an injury follows unsafe behavior. The "It can't happen to me" attitude is usually supported or rewarded by our actual experiences. Risk-taking is rarely punished with an injury or even a near hit, instead it's consistently rewarded with convenience, comfort, or time saved. This creates something of a vicious cycle. The rewards of risky behavior mean you're likely to take more chances. As you gain experience at work you often master dangerous shortcuts. Since these at-risk behaviors are not followed by a near hit or injury, they remain unpunished, and they persist.

This basic principle of human nature reinforced throughout our lives runs counter to the safety efforts of individuals, groups, organizations, and communities. It explains why promoting safety and health is the most difficult ongoing challenge at work. The reality is that injuries really do happen to the "other guy."

Dimensions of Human Nature

I always categorize factors contributing to a work injury into three areas: (1) environment factors, (2) person factors, and (3) behavior factors. This is the "Safety Triad" introduced in Chapter 2. The most common reaction to an injury is to correct something about the environment— modify or fix equipment, tools, housekeeping, or an environmental hazard.

Often the accident report includes some mention of personal factors, like the employee's knowledge, skills, ability, intelligence, motives, or personality. These factors are typically translated into general recommendations, such as "The employee will be disciplined," or "The employee will be retrained."

This kind of vague attention to critical human aspects of a work injury shows how frustrating and difficult it is to deal with the psychology of safety—

the person and behavior sides of the Safety Triad. The human factors contributing to injury are indeed complex, often unpredictable and uncontrollable. This justifies my conclusion that not *all* injuries can be prevented.

The acronym **BASIC ID** reflects the complexity and uncontrollability of human nature. As depicted in Figure 4.1, each letter represents one of seven human dimensions. A simple scenario underscores the need in safety to understand various human factors:

B EHAVIOR

A TTITUDE

S ENSATION

I MAGERY

C OGNITION

I NTERPERSONAL

D RUGS

Figure 4.1. The acronym BASIC ID reflects the complexity of people and potential contributions to injury.

Dave, an experienced and skilled crafts-man, works rapidly to make an equipment adjustment while the machinery continues to operate. As he works, production-line employees watch and wait to resume their work. Dave realizes all too well that the sooner he finishes his task, the sooner his coworkers can resume quality production. So he doesn't shut down and lock out the equipment power. After all, he's adjusted this equipment numerous times before without locking out and he's never gotten injured.

A morning argument with his teenage daughter pervades Dave's thoughts as he works, and suddenly he experiences a near hit. His late timing nearly results in his hand being crushed in a pinch point.

Removing his hand just in time, Dave feels weak in his knees and begins to perspire. This stress reaction is accompanied by a vivid image of a crushed right hand. After gathering his composure, Dave walks to the switch panel, shuts down and locks out the power, then lights up a cigarette. He thinks about this scary event for the rest of the day, and talks about the near hit to fellow workers during his breaks.

This brief episode illustrates each of the psychological dimensions represented by BASIC ID (see Figure 4.1), and demonstrates the complexity of human activity.

Behavior is illustrated by observable actions such as adjusting equipment, pulling a hand away from the moving machinery, lighting up a cigarette, and talking to coworkers.

Dave's **attitude** about work was fairly neutral at the start of the day, but immediately following his near hit he felt a rush of emotion. His attitude toward "energy control and power lock out" changed dramatically, and his commitment to locking out increased after relating his near hit to friends.

Sensation is evidenced by Dave's dependence on visual acuity, hand-eye coordination, and a keen sense of timing when adjusting the machinery. His ability to react quickly to the dangerous situation prevented severe pain and potential loss of valuable touch sensation.

Imagery occurred after the near hit when Dave visualized a crushed hand in his mind's eye, and this contributed to the significance and distress of the incident. Dave will probably experience this mental image periodically for some time to come. And this will motivate him to perform appropriate lock-out procedures, at least for the immediate future.

Cognition or "mental speech" about the morning argument with his daughter may have contributed to the timing error that resulted in the near hit. Dave will probably remind himself of this episode in the future, and these cognitions may help trigger proper lock-out behavior.

Interpersonal refers to the other people in Dave's life who contributed to his near hit, and will be influential in determining whether he initiates and maintains appropriate lock-out practices.

For example, it was the interpersonal discussion with his daughter that occupied his thoughts or cognitions before the near hit. The presence of production-line workers influenced Dave through subtle peer pressure to quickly adjust equipment without lock-out practices. These onlookers may have distracted Dave from the task, or they could have motivated him to show off his adjustment skills. After Dave's near hit, his interpersonal discussions were therapeutic, helping him relieve his distress and increase his personal commitment to occupational safety.

Drugs in the form of caffeine from morning coffee may have contributed to Dave's timing error. The extra cigarettes Dave smokes as a "natural" reaction to distress also had physiological consequences, which could have been reflected in Dave's subsequent behavior, attitude, sensation, or cognition. Dave's lesson shows how human nature interacts with environmental factors to cause at-risk work practices, near hits, and sometimes personal injuries. It's relatively easy to control the environmental factors. And, as I'll explain in Part Three on behavior-based safety, it's feasible to measure and control the behavior factors. However, the complex person factors, described by the BASIC ID acronym, are quite elusive. The field of psychology provides insights here, and this information can benefit occupational safety and health programs.

Let's further discuss person-based dimensions that make safety achievement so challenging.

COGNITIVE FAILURES

"All injuries are preventable."

I've heard this said so many times that it seems to be a slogan or personal affirmation that safety pros use to keep themselves motivated. I suspect some readers will resist any challenge to this ideal. I certainly appreciate their optimism, and there's no harm if such perfectionism is kept to oneself. But sharing this belief with others can actually inhibit achieving a Total Safety Culture.

You see, if a common workplace slogan declares all injuries preventable, workers may be reluctant to admit they were injured, or had a near hit. After all, if all injuries are preventable and I have an injury, I must be a real "jerk" for getting hurt.

Combine this slogan with a *goal* of zero injuries and a *reward* for not having an injury and human nature will dictate covering up an injury if possible. Also, as I'll discuss in the next chapter, claiming that all injuries are preventable can reduce the perceived risk of the situation. This can create the notion that "it won't happen to me," a perception that can increase the probability of at-risk behavior and an eventual injury.

Eliminate the "all injuries are preventable" slogan from your safety discussions. The most important reason to drop it is that most people don't believe it, anyway. They have been in situations where all the factors contributing to the near hit or injury could not have been anticipated, controlled, or prevented. The most uncontrollable factors are the person-based or internal subjective dimensions of people. Consider, for example, the role of cognitive failures.

Just what is a "cognitive failure"? I liken it to a "brain cramp." And research by Broadbent *et al,* 1982, demonstrated that people who report greater frequency of "cognitive failures" are more likely to experience an injury. The items listed in Figure 4.2 were used by Professor Broadbent and his associates to measure cognitive failures. Respondents were merely asked to indicate on a five-point scale the extent to which they agreed with each statement (from "strongly disagree" to "strongly agree"). Professor Broadbent's measurement instrument offers an operational definition for this person dimension that apparently influences injury frequency.

Scientific protocol won't allow us to conclude from the research by Broadbent *et al* that cognitive failures cause injury, but based on personal experience it sure seems reasonable to interpret a cause-and-effect relationship. It's likely every reader has experienced one or more of the "brain cramps" listed in Figure 4.2. Surely you have walked into a room to get something and forgotten why you were there. And how often have you left home for work more than once in a single morning because you forgot something? The same sort of breakdown in cognitive functioning can cause an injury. Does Figure 4.3 reflect potential reality?

Being actively aware of possible cognitive failures (listed in Figure 4.2) can help us develop strategies to decrease how often we experience them. For

- I sometimes forget why I went from one part of the house to another.
- I often fail to notice signposts on the road.
- I sometimes bump into things or people.
- I often forget whether I've turned off a light or the coffeepot, or locked the door.
- I sometimes forget which way to turn on a road I know well but rarely use.
- I sometimes fail to see what I want in a supermarket (although it's right there).
- I often forget where I put something like a newspaper or a book.
- I often daydream when I ought to be listening to something.
- At home, I often start doing one thing and get distracted by something else (unintentionally).
- I sometimes forget what I came to the store to buy.
- I often drop things.
- I often find myself putting the wrong things in the wrong place when I'm done with them—like putting milk in the cereal cupboard.
- I frequently confuse right and left when giving directions.

Figure 4.2. Items from the survey developed by Broadbent *et al* (1982) to measure cognitive failures.

example, reducing fatigue and stress (discussed in Chapter 6) can make a marked difference. But I doubt that any personal strategy can eliminate all cognitive failures. Furthermore, it's unlikely you can have much impact on the cognitive failures of others.

So, if you believe as I do that "brain cramps" can cause personal injury, and that it's impossible to control completely these sudden, unexpected mental lapses, then it's illogical to presume that all injuries are preventable. There's a benefit to realizing that personal factors like cognitive failures are extremely hard to control—it should make us less likely to take

Figure 4.3. A cognitive failure or "brain cramp" can cause a workplace injury.

risk for granted, even in familiar situations. We should be motivated to take greater precautions or to work more defensively. But this isn't easy, and interpersonal factors are one reason why. Now let's look more closely at the second *I* of BASIC ID.

INTERPERSONAL FACTORS

Our interpersonal relationships dramatically influence our thoughts, attitudes, and actions. How much of your time each day goes into efforts to gain the approval of others? Of course, we sometimes attempt to avoid the disapproval of others, be they a parent, spouse, work supervisor, or department head. As discussed in Chapter 3, we don't feel as good—or as *free*—when working to avoid failure or disapproval as when working to achieve success or approval. In both cases, other people are the cause of our motivation and behavior.

Social psychologists study systematically how we all influence each other's attitudes and behaviors. I want to discuss two reliable phenomena from social psychology research that strongly affect our daily activities. These phenomena are no doubt responsible for many at-risk behaviors and injuries, yet they are often overlooked.

PEER PRESSURE

Research conducted by Solomon Asch and associates in the mid-1950s found more than one out of three intelligent and well-intentioned college students were willing to publicly deny reality in order to follow the obviously inaccurate judgments of their peers. Dr. Asch's classic studies of conformity (1956, 1958) involved six to nine individuals sitting around a table judging which of three comparison lines was the same length as the standard.

All but the last individual to voice an opinion were research associates posing as subjects. Sometimes the research associates uniformly gave obviously incorrect judgments. The last person to decide was the real subject of the experiment. About one-third of the time the subject denied the obvious truth in order to go along with the group consensus.

The phenomenon of social conformity is certainly not new to any reader. We see examples of it every day, from the clothes people wear to how they communicate both verbally and in writing. We can't overlook the power of conformity in influencing at-risk behavior. Plus, we've learned that peer pressure increases when more people are involved, and when the group members are seen as relatively competent or experienced. Can you relate, for example, to the workplace conformity situation illustrated in Figure 4.4?

It's important to remember, though, that one dissenter—a leader willing to ignore peer pressure and do the right thing—is often enough to prevent another person from succumbing to potentially dangerous conformity at work.

Figure 4.4. Social conformity can be risky.

POWER OF AUTHORITY

Imagine you're among nearly 1000 participants in one of professor Stanley Milgram's 20 obedience studies at Yale University in the 1960s. You and another individual are led to a laboratory to participate in a human learning experiment. First, you draw slips of paper out of a hat to determine randomly who will be the "teacher" and who the "learner."

You get to be the teacher; the learner is taken to an adjacent room and strapped to a chair wired through the wall to an electric shock machine containing 30 switches with labels ranging from "15 volts—light shock" to "450 volts—severe shock." You sit behind this shock generator and are instructed to punish the learner for errors in the learning task by delivering brief electric shocks, starting with the 15-volt switch and moving up to the next higher voltage with each of the learner's errors. The scenario is depicted in Figure 4.5.

Complying with the experimenter's instructions, you hear the learner moan as you flick the third, fourth, and fifth switches. When flicking the eighth switch (labeled "120 volts"), the learner screams "These shocks are painful"; and when the tenth switch is activated the learner shouts "Get me out of here!"

At this point, you might think about stopping, but the experimenter prompts you, "Please continue— the experiment requires that you continue."

Figure 4.5. Subjects readily gave electric shocks to peers.

Increasing the shock intensity with each of the learner's errors, you reach the 330-volt level. Now you hear shrieks of pain—the learner pounds on the wall, then becomes silent. Still, the experimenter urges you to flick the 450-volt switch when the learner fails to respond to the next question.

At what point will you refuse to obey the instructions? Milgram asked this question to a group of people, including 40 psychiatrists, before conducting the experiment. They thought the sadistic game would stop soon after the learner indicated the shock was painful. So Milgram and his associates (1963) were surprised that 65 percent of his actual subjects, ranging in age from 20 to 50, went along with the experimenter's request right up to the last 450-volt switch.

Why did they keep following along? Did they figure out that the learner was a confederate of the experimenter and did not really receive the shocks? Did they realize they were being deceived in order to test their obedience?

No, the subjects sweated, trembled, and bit their lips when giving the shocks. Some laughed nervously. Others openly questioned the instructions. But most did as they were told.

Milgram and associates learned more about the power of authority in further studies. Full obedience exceeded 65 percent, with as many as 93 percent flicking the highest shock switch, when:

> The authority figure—the one giving the orders—was in the room with the subject;
>
> The authority was supported by a prestigious institution, such as Yale University;
>
> The shocks were given by a group of "teachers" in disguise and remaining anonymous;
>
> There was no evidence of resistance—no other subject was observed disobeying the experimenter;
>
> The victim was depersonalized or distanced from the subject in another room.

Stanley Milgram drew this lesson from the research: "Ordinary people, simply doing their jobs, and without any particular hostility on their part, can become agents in a terrible destructive process" (Milgram, 1974, p.6).

Let's apply this research to the workplace. As a result of social obedience or social conformity, people might perform risky acts or overlook obvious safety hazards, and put themselves and others in danger. To say, "I was just following orders" reflects the obedience phenomenon, and "Everyone else does it" implies social conformity or peer pressure.

To achieve a Total Safety Culture, we need to realize the power of these two interpersonal factors. Interventions capable of overcoming peer pressure and blind obedience are detailed in Part Three. What I want to stress at this point is the vital role of leadership. One person can make a difference—decreasing both destructive conformity and obedience—by deviating from the norm and setting a safe example. And when a critical mass of individuals boards the "safety bandwagon," you get constructive conformity and obedience that supports a Total Safety Culture.

In Conclusion

We need to understand a problem as completely as possible and from many perspectives before we can solve it. In this chapter, we explored dimensions of the safety problem by considering the complexity of people. I attempted to convince you that human nature does not usually support safety. The natural relationships between behavior and its motivating consequences usually result in some form of convenient, time-saving—and risky—behavior. Consequently, to achieve a Total Safety Culture, you should prepare for an ongoing fight with human nature.

Human barriers to safety are represented by a popular acronym from clinical psychology (BASIC ID). The *C* (cognitions) and second *I* (interpersonal) dimensions of this acronym in particular explain the special challenges of achieving a Total Safety Culture. The phenomenon of cognitive failures shows the shallowness—in fact the potential danger—of the popular safety slogan, "All injuries are preventable." Conformity and obedience, two powerful phenomena from social psychological research, further help us understand the individual, group, and system factors responsible for at-risk behavior and injury.

The human barriers to safety discussed here should lead us to be more defensive and alert in hazardous environments. They also show how difficult it is to find the root cause, if indeed there is such a thing, of a near hit or injury. Another psychological challenge to safety is explored in the next chapter when we discuss the *S* (sensation) of BASIC ID.

Chapter 5

Sensation, Perception, and Perceived Risk

It's critically important to understand that perceptions of risk vary among individuals. We can't dramatically improve safety until people increase their perception of risk in various situations, and reduce their overall tolerance for risk.

In this chapter we'll explore the notion of selective sensation or perception, and then relate the concept to perceived risk and injury control. Several factors will be discussed that impact whether employees react to workplace hazards with alarm, apathy, or something in between. Taken together, these factors shape personal perceptions of risk and illustrate why the job of improving safety is so challenging.

The **S** of the BASIC ID acronym introduced in Chapter 4 refers to **sensation**—a human dimension that influences our thinking, attitudes, emotions, and behavior. In grade school we learned there are five basic senses we use daily to experience our world (we see, hear, smell, taste, and touch). Later we learned that our senses do not take in all of the information available in our immediate surroundings. Instead, we intentionally and unintentionally tune in and tune out certain features of our environment; thus, some potential experiences are never realized.

Selective Sensation or Perception

At the 1994 Professional Development Conference of the American Society of Safety Engineers (ASSE), the following instructions were printed in half of the 40-page handouts distributed to the audience of more than 350 individuals at the start of my two-hour presentation.

Figure 5.1. Ambiguous drawing illustrates selective sensation. Please read instructions, then look at the drawing for five seconds. Afterward, answer the five questions.

You are going to look briefly at a picture and then answer some questions about it. The picture is a rough sketch of a poster of a couple at a costume ball. Do not dwell on the picture. Look at it only long enough to "take it all in" at once. After this, you will answer "yes" or "no" to a series of questions.

After the participants read the instructions, I presented the illustration depicted in Figure 5.1 for about five seconds. If you would like to experience the biased visual sensation (or perception) demonstrated to the ASSE audience, please read the instructions given above and then look at Figure 5.1 for approximately five seconds.

Now, answer the questions that I asked my ASSE audience:

1. Did you see a man in the picture?

2. Did you see a woman in the picture?

3. Did you see a animal in the picture? If so, what kind of animal did you see?

4. What other details did you detect in the brief exposure to the drawing?__ A woman's purse?__ A man's cane?__ A trainer's whip?__ A fish?__ A ball?__ A curtain?__ A tent?

Practically everyone in the audience raised a hand to answer "yes" to the first question, and I suspect you also see a man in Figure 5.1. But only about half of the audience acknowledged seeing a woman in the drawing, and many said they had seen an animal.

When I asked what type of animal, the common response heard across the room was "seal." This drew many laughs, and the laughter got louder when I asked what else was quickly perceived in the illustration.

Several people saw a woman's purse and a man's cane; others said they had seen a trainer's whip, a fish, and a beach ball. Some remembered seeing a curtain. Others saw part of a circus tent. What did you see in Figure 5.1?

"What's going on here?" I asked the ASSE audience. Why are we getting these diverse reactions to one simple picture? Some people speculated about environmental factors in the seminar room, including lighting, spatial orientation and visual distance from the presentation screen. Others thought that individual differences, including gender, age, occupation, and personal experiences even "last night," could be responsible. Finally, someone asked whether the instructions printed in their handouts could have influenced the different perceptions. This was in fact the case.

Every handout included the same exact instructions except for a few words. Half included all the words given above; the rest had the words "trained seal act" substituted for "couple at a costume ball." This was enough to make a marked difference in perceptions. Perhaps this makes perfect sense to you. Critical words in the instructions created expectations for a particular visual experience. I had set up my audience. Was your perception of Figure 5.1 influenced by this "set up"?

BIASED BY CONTEXT

Now take a look at Figure 5.2. I suspect you have no difficulty reading the sentence as "The cat sat by the door," even though the symbol for "H" is exactly the same as the symbol for "A." It's a matter of context. The symbol was positioned in a way that influenced your labeling (or encoding) of the symbol. Likewise, the context or environmental surroundings in our visual field influence how we see particular stimuli.

The same is true for our other senses—hearing, smelling, tasting, and touching. How we experience food, which involves the sensations of smell,

THE CAT SAT
BY THE DOOR

Figure 5.2. The context or circumstances surrounding a stimulus can influence how we perceive it.

47

Figure 5.3. The environmental context influences personal perception of the man in uniform.

touch, and taste, can be dramatically influenced by the atmosphere in which it is served. This is a basic rule of the restaurant business. Of course other factors also bias food sensations, including hunger and past experiences with the same and similar food.

Take a look at Figure 5.3. What label do you give the man in the drawing on the left? The setting or context certainly influences your decision. The sign, keys, and uniform are cues that the man is probably a doorman. The environmental context in the drawing on the right leads to a different perception and label for the same person. Here he is a policeman enforcing a safety policy.

Now let's take our discussion of perception and apply it to the workplace. Here, perceptions of people can be shaped by equipment, housekeeping, job titles, and work attire. In fact, our own job title or work assignment can influence perceptions of ourselves, as well as affect our perceptions of others. This can dramatically influence how we interact with others, if indeed we choose to interact at all.

It's important to recognize this contextual bias. Pick out someone you communicate with at work, and think how your relationship would be different in another setting. Would you still feel superior or inferior? Also, as depicted in Figure 5.4, the work setting has a way of turning individuals into numbers, depersonalizing them. This impression certainly can be misleading, and might cause you to overlook someone's potential. In another setting, the same individual might feel empowered and be perceived as a leader.

BIASED BY OUR PAST

Perhaps every reader realizes that our past experiences influence our present perceptions. In Chapter 3, we considered shifts in methods and percep-

Figue 5.4. The environmental context influences personal perception and behavior.

tions needed to achieve a Total Safety Culture. When I give workshops on paradigm shifts, someone invariably expresses concern about resistance. "He (or she) keeps playing old tapes and is not open to new ideas," is a common refrain. Past experiences are biasing present perceptions. Actually, there is a long trail of inter-twined factors here. Past experiences filter through a personal evaluation process that is influenced by person factors, including many past perceived experiences. The cumulative collection of these previous perceived experiences biases every new experience, and makes it indeed difficult to "teach an old dog new tricks."

Some participants arrive at my seminars and workshops with a "closed mind" and a "have to be here" attitude. Others start with an "open mind" and an "opportunity to learn" outlook. This is another example of the power of personal perception—how much one learns at these seminars depends on perceptions going in.

Perhaps you'll find it worthwhile to copy Figure 5.5 and use it for a group demonstration. You can show how current impressions are affected by prior perceptions by asking participants to call out what they see as you reveal each drawing. The drawings must be uncovered in a particular order. Show the top row of pictures first, revealing each successive one from left to right. The last picture will probably be identified as the face of an elderly man.

With the top row covered, then show the successive animal pictures of the second row. Now the last picture will likely be identified as a rat or mouse. Even after knowing the purpose of the demonstration, you can view serially the row pictures in Figure 5.5 and see how your perception of the last drawing changes depending on whether you previously looked at human faces or animals.

Figure 5.5. Prior perceptual experience influences current perception. (Adapted from Bugelski & Alimpay, 1961)

Now I'd like you to read the sentence given in Figure 5.6 with the intent of understanding what it means. The sentence might seem to make little sense, but treat it as a sentence in a memo you have received from a colleague or supervisor. Some of those memos seem meaningless, too. Your past experience at reading memos, as well as your mood at the time, can influence how you perceive and react to a memo.

After reading the sentence in Figure 5.6, go back and quickly count the number of times the letter *F* occurs in the sentence. Record your answer. When I show this sentence to workshop participants and ask the same question, most will answer "three." A few will shout out "six," usually because they have seen the demonstration before. "Six" is indeed the correct answer, but even after knowing this, a number of people cannot find more than three *F*s in the sentence. Why?

When I was first introduced to this exercise many years ago, I showed the sentence to my two young daughters, and they both found six *F*s immediately. Karly was in kindergarten and Krista was a second grader. Neither could understand the words.

FINISHED FILES ARE THE RESULT OF YEARS OF SCIENTIFIC STUDY COMBINED WITH THE EXPERIENCE OF MANY YEARS.

Figure 5.6. Past experience can teach us to overlook details.

My wife had the same difficulty as I, and could only find three. I remember looking at the sentence again and again, trying to find six *F*s but to no avail. My past experience at speed reading had conditioned me to simply overlook small, unessential words like "of." My history had biased my perception. I bet you had a similar experience, unless you had seen this demonstration before. If so, then *that* experience biased your current perception of Figure 5.6.

If you need more evidence that past experience influences current perceptions, take a look at the woman in Figure 5.7.

Notice anything strange, other than that the picture is upside-down? Is this face relatively attractive, or at least normal? Now turn the book upside-down and view the

Figure 5.7. Viewing the face from a different orientation (by turning the book upside-down) will influence a different perception.

woman's face right-side-up. Has your perception changed? Why didn't you notice her awkward (actually ugly) mouth when the picture was upside-down? Perhaps both context and prior experience (or learning) biased your initial perception. I bet this perceptual bias will persist even after you realize the cause of the distortion, and after viewing the face several times in both positions. A biased perception can be difficult to correct. It's not easy to fight human nature.

RELEVANCE TO ACHIEVING A TOTAL SAFETY CULTURE

Is the relevance of this discussion to occupational safety and health obvious? Perhaps by understanding factors that lead to diverse perceptions we can become more tolerant of individuals who don't appear to share our opinion or viewpoint. Perhaps the person factors discussed here increase your appreciation and respect for diversity and support the basic need to actively listen. "Seek first to understand, before being understood" is Stephen Covey's fifth habit for highly effective people (Covey, 1989, p.235).

It's also possible that this discussion and the exercises on personal perception have reduced your tendency to blame individuals for an injury or to look for a single root cause of an undesirable incident. Before we react to an incident or injury with our own viewpoint, recommendation, or corrective action, we need to ask others about their perceptions.

Hopefully, I have not reduced your optimism toward achieving a Total Safety Culture. But maybe I've alerted you to challenges not previously considered. If I haven't convinced you yet to stop claiming "All injuries are preventable," the next section should do the trick.

Perceived Risk

People are generally underwhelmed or unimpressed by risks or safety hazards at work. Why? Our experiences on the job lead us to perceive a relatively low level of risk. This is strange. After all, it's quite probable someone will eventually be hurt on the job when you factor in the number of hours workers are exposed to various hazards.

In Chapter 4, I discussed one major reason for low perceptions of risk in the workplace. It's elemental, really—we usually get away with risky behavior. As each day goes by without receiving an injury, or even a near hit, we become more accepting of the common belief, "It's not going to happen to me." Now let's further explore why we're generally not impressed by safety hazards at work.

REAL VERSUS PERCEIVED RISK

The real risk associated with a particular hazard or behavior is determined by the magnitude of loss if a mishap occurs, and the probability that the loss or accident will indeed occur. For example, the risk that comes from driving during any one trip can be estimated by calculating the probability of a vehicle crash on one trip and multiplying this value by the magnitude of injury from a crash. Of course, the injury potential or mortality rate from a vehicle crash is influenced by many other factors, including size of vehicle(s) involved, speed of vehicle(s), road conditions, and whether the vehicle occupants were using safety belts.

On any single trip the chance of a vehicle crash is minuscule; but in a lifetime of driving the probability is quite high, varying from .30 to .50 depending upon factors such as geographic location, trip frequency and duration, and characteristics of the driver such as age, gender, reaction time, or mental state. Obviously, the risk of driving an automobile is difficult to assess, although it has been estimated that 55 percent of all fatalities and 65 percent of all injuries would have been prevented if a combination shoulder and lap belt had been used.

Estimating the risk of injury from working with certain equipment is even more difficult to determine, because work situations vary so dramatically.

Plus the risk can be eliminated completely by the use of appropriate protective clothing or equipment. Still, many people don't appreciate the value of using personal protective equipment or following safe operating procedures. Their perception of risk is generally much lower than actual risk. This thinking pervades society. Automobile accidents are the nation's leading cause of lost productivity, greater than AIDS, cancer, and heart disease. But how many of us take driving for granted?

The risk of a fatality from driving a vehicle or working in a factory is much higher than from the environmental contamination of radiation, asbestos, or industrial chemicals. Yet look at the protests over asbestos in schools and neighborhood chemical plants.

Researchers of risk communication have found that various characteristics of a hazard, irrelevant to actual risk, influence people's perceptions. It's important to consider these characteristics, because behavior is determined by perceived rather than actual risk.

Figure 5.8 shows factors that influence our risk perceptions. It is derived from research by Drs. Peter Sandman (1991), Paul Slovic (1991), and their colleagues. The factors listed on the left reduce perceptions of risk and are typically associated with the workplace. The opposing factors in the right-hand column have been found to increase risk perception, and these are not usually experienced in the work setting. As a consequence, our perception of risk on the job is not as high as it should be, and therefore we don't work as defensively as we should. Discussing some of these factors will reveal strategies for increasing our own and others' perception of risk in certain situations.

LOWER RISK	HIGHER RISK
• exposure is voluntary	• exposure is mandatory
• hazard is familiar	• hazard is unusual
• hazard is forgettable	• hazard is memorable
• hazard is cumulative	• hazard is catastrophic
• collective statistics	• individual statistics
• hazard is understood	• hazard is unknown
• hazard is controllable	• hazard is uncontrollable
• hazard affects anyone	• hazard affects vulnerable people
• preventable	• only reducible
• consequential	• inconsequential

Figure 5.8. Factors which influence perceived risk. Those in the left-hand column *reduce* risk perception and are generally associated with the workplace. (Adapted from Sandman, 1991)

THE POWER OF CHOICE

Hazards we choose to experience (like driving, skiing, and working) are seen as less risky than ones we feel forced to endure (like food preservatives, environmental pollution, and earthquakes). Of course, the perception of choice is also subjective, varying dramatically among individuals. For example, people who feel they have the freedom to pull up stakes and move whenever they want would likely perceive less risk from a nearby nuclear plant or seismic fault. Likewise, employees who feel they have their pick of places to work generally perceive less risk in a work environment. They are typically more motivated and less distressed. In the next chapter, I'll discuss relationships between perceived choice, stress, and distress.

FAMILIARITY BREEDS COMPLACENCY

Familiarity is probably a more powerful determinant of perceived risk than choice. The more we know about a risk, the less it threatens us. Remember how attentive you were when first learning to drive, or when you were first introduced to the equipment in your workplace? It wasn't long before you lowered your perceptions of risk, and changed your behavior accordingly. When driving, for example, most of us quickly shifted from two hands on the wheel and no distractions to steering with one hand while turning up the radio and carrying on a conversation.

It's so easy to tune out the familiar hazards of the workplace. Safety professionals respond by constantly reminding employees of risks with a steady stream of memos, newsletters, safety meetings, and signs. Still, these efforts can't compete with the impact of unusual, catastrophic, and memorable events broadcast by the media and dramatized on television and in the movies. Publicity of memorable injuries, like those suffered by John Wayne Bobbitt and Nancy Kerrigan in 1994, influences misperception of actual risk.

SYMPATHY FOR VICTIMS

Many people feel sympathy for victims of a publicized incident, even vividly visualizing the injury as if it happened to them. Personalizing these experiences increases perceived risk. At work, employees show much more attention and concern for hazards when injuries or near hits are discussed by the coworkers who experienced them, compared to a presentation of statistics. The average person can't relate to group numbers. But there's power in personal stories. I have met many people over the years who accepted individual accounts in lieu of convincing statistics: "The police officer told Uncle Jake he would have been killed if he had been buckled up"; "Aunt Martha is 91 years old and still smokes two packs of cigarettes a day."

This suggests that we should shift the focus of safety meetings away from statistics, emphasizing instead the human element of safety. Safety talks and intervention strategies should center on individual experiences rather than numbers. This might be easier said than done. Encouraging victims to come forward with their stories is often stifled by management systems in many companies that seem to value fault finding over fact finding, piecemeal rather than system approaches to injury investigation, and enforcement and discipline more than recognition and rewards to influence on-the-job behaviors.

UNDERSTOOD AND CONTROLLABLE HAZARDS

Hazards we can explain and control cause much less alarm than hazards that are not understood, and thus are perceived as uncontrollable. This points up a problem with many employee safety education and training programs. Workplace hazards are explained in a way that creates the impression they can be controlled. Indeed, safety professionals often state a vision or goal of "zero injuries," implying complete control over the factors that cause injuries. This actually lowers perceived risk by convincing people the causes of occupational injuries are understood and controllable.

Perhaps it would be better for safety leaders to admit and publicize that only two of the three types of factors contributing to workplace injuries can be managed effectively—environmental/equipment factors and work behaviors. As I've already discussed in preceding chapters, the mysterious inside, unobservable, and subjective world of people dramatically influences the risk of personal injury. These attitudes, expectancies, perceptions, and personality characteristics cannot be measured, managed, or controlled reliably. Internal human factors make it impossible to prevent all injuries. By discussing the complexity of people and their integral contribution to most workplace hazards and injuries, you can increase both the perceived value of ongoing safety interventions and the belief that a Total Safety Culture requires total commitment and involvement of all concerned.

ACCEPTABLE CONSEQUENCES

We're less likely to feel threatened by risk-taking or a risk exposure that has its own rewards. But if few benefits are perceived by an at-risk behavior or environmental condition, outrage—or heightened perceived risk—is likely to be the reaction, along with a concerted effort to prevent or curtail the risk.

Some people, for example, perceive guns, cigarettes, and alcohol as having limited benefit and thus lobby to restrict or eliminate these societal hazards. The availability of and exposure to these hazards will continue, though, as long as a significant number of individuals perceive that the risk benefits outweigh the risk costs. You can see how cost-benefit analyses are easily politicized by subjective perceptions of risk.

The benefits of risky work behaviors are obvious. You might feel cooler and more comfortable not wearing a respirator. It's convenient—you can talk easier. Feeling relieved, you might actually be more productive. The costs of not wearing the mask might be abstract and delayed (if the exposure is not immediately life-threatening). Statistics might point out a chance of getting a lung disease, which won't surface for decades, if ever. Decisions about risk-taking are made every day by workers. By playing the odds and shooting for short-term gains, risky work practices are often accepted and not perceived to be as dangerous as they actually are.

SENSE OF FAIRNESS

Most people believe in a just and fair world. "What goes around comes around." We believe people generally get what they deserve. When people receive benefits like increased productivity from their risky behavior, the out-rage, public attention, or perceived risk is relatively low. On the other hand, when hazards or injuries seem unfair, as when a child is molested or inflicted with a deadly disease, special attention is given. This increased attention results in more perceived risk.

This makes it relatively easy to obtain contributions or voluntary assis-tance for programs that target vulnerable populations, like learning-disabled children. Workplace injuries, though, are perceived as fair. They are indis-criminately distributed among employees who take risks, and these people deserve what they get. This perception or attitude naturally lowers the outrage we feel when someone gets injured on the job. Lower outrage translates into lower perceived risk.

Risk Compensation

A discussion of risk perception would not be complete without examining one of the most controversial concepts in the field of safety. In recent years, it has been given different labels, including *risk homeostasis, risk or danger compensation, risk-offsetting behavior,* and *perverse compensation.* Whatever the name, the basic idea is quite simple and straightforward. People are pre-sumed to adjust their behavior to compensate for changes in perceived risk. If a job is made safer with machine guards or the use of personal protective equipment, workers might reduce their perception of risk and thus perform more recklessly. For example, if the individual depicted in Figure 5.9 is taking a risk due to the perceived security of fall protection, we would have support for risk compensation theory.

The notion of taking more risks to compensate for lower risk perception certainly seems intuitive. I bet every reader has experienced this phenome-non. I clearly remember taking more risks after donning a standard high school football uniform. With helmet and shoulder pads, I'd willingly throw

Figure 5.9. Personal protective equipment can reduce the perception of risk.

Figure 5.10. Back belts can give a false sense of security.

my body in the path of another player or leap to catch a pass. I didn't perform these behaviors until perceiving security from the personal protective equipment (PPE).

Figure 5.10 depicts a workplace situation quite analogous to my teenage experiences on the football field. If the use of a back belt leads to employees lifting heavier loads, then the potential protection from this device could be offset by greater risk taking. The protective device could give a false sense of security and reduce one's perception of being vulnerable to back injury. The result could be more frequent and heavier lifting, and greater probability of injury. This is why back belt suppliers emphasize the need for training and education in the use of belts.

IMPLICATIONS OF RISK COMPENSATION.

I'm convinced from personal experience and reading the research literature that risk compensation is a real phenomenon. What does this mean for injury prevention? Professor Wilde (1994) says it means safety excellence cannot be achieved through top-down rules and enforcement. Some people only follow the rules when they are supervised, and might take greater risks when they can get away with it. As the title of his book *Target Risk* indicates, Professor Wilde advocates that safety interventions need to lower the level of risk people are willing to tolerate. This requires a change in values. Wilde (1994) asserts that improvements in safety cannot be "achieved by interventions in the form of training, engineering or enforcement" (p. 213). "The extent of risk taking with respect to safety and health in a given society, therefore, <u>ultimately depends on values that prevail in that society, and not on the available technology</u>" (Wilde, 1994, p. 223).

I hope it's obvious that Wilde's position is consistent with the theme of this text. When people understand and accept the paradigm shifts needed for a Total Safety Culture (see Chapter 3), they are on track to reducing their tolerance for risk. Next they need to believe in the vision of a Total Safety Culture and buy into the mission of achieving it. Then they need to understand and accept the procedures that can achieve this vision. These methods are explained in Part Three of this text. Through a continuous process of applying the right procedures, the work force will feel empowered to actively care for a Total Safety Culture. Finally, they will come to treat safety as a value rather than a priority. I discuss these concepts more fully in Part Four.

In Conclusion

This chapter explored the concept of selective sensation or perception, and related it to perceived risk and injury control. Visual exercises illustrated the impact of past experience and contextual cues on present perception. This allows us to appreciate diversity and realize the value of actively listening during personal interaction. We need to work diligently to understand the perceptions of others before we impulsively jump to conclusions or attempt to exert our influence.

<u>We must realize that perceptions of risk vary dramatically among individuals. And we can't improve safety unless people increase their perception of, and reduce their tolerance for, risk.</u> Changes in risk perception and acceptance will occur when individuals get involved in achieving a Total Safety Culture with the principles and procedures discussed in this book.

Several factors were discussed in this chapter that affect whether employees react to workplace hazards with alarm, apathy, or something in between. Taken together, these factors shape personal perceptions of risk and illustrate why the job of improving safety is so daunting. This justifies more resources for safety and health programs, and intervention plans to motivate continual employee involvement. We will get to a discussion of these interventions in Part Three. But before discussing strategies to fix the problem, we need to understand how stress, distress, and personal attributions contribute to the problem. That's our topic for the next chapter.

Chapter 6

Stress versus Distress

Stressors can contribute to a near hit or injury; they are barriers to a Total Safety Culture. However, stressors can provoke positive stress rather than negative distress—which can lead to constructive problem solving rather than destructive, at-risk behavior. This chapter explains the important distinction between stress and distress, and defines factors which determine the occurrence of one or the other.

The concept of "attribution" is reviewed as a process we use to reduce personal distress. This is one way accident victims shift blame and distort follow-up investigations. I show, however, how this self-serving bias can actually be used to benefit safety efforts.

Judy was tired and worried. She had just left her six-year-old son at her sister's house with instructions for her to take him to Dr. Slayton's office for a 10:30 a.m. appointment. She had been up much of the night with Robbie, attempting to comfort him. With tears in his eyes, he had complained of a "hurt" in his stomach. This was the third night his cough had periodically awakened her, but last night Robbie's cough was deeper, seemingly coming from his lungs.

Judy arrived at her work station a little later than normal and found it more messy than usual. Grumbling under her breath that the night shift had been "careless, sloppy, and thoughtless," she downed her usual cup of coffee and waited for the production line to crank up, rather than cleaning up the work area. After all, it wasn't her mess. The graveyard shift is not nearly as busy as the day shift. How could they be so sloppy and inconsiderate?

Judy was ready to start her inspection and sorting when she noticed the "load cart" was misaligned. She inserted a wooden handle in the bracket and pulled hard to jerk the cart into place. Suddenly the handle broke, and Judy fell backward against the control panel. Fortunately, she wasn't hurt, and the only damage was the broken handle. Judy discarded it, inserted another one, and put the cart in place.

During lunch Judy called the doctor's office and learned that her son had the flu, and would be fine in a day or two. She completed the day in a much better mood, and without incident.

At the end of her shift, Judy filled out a near-hit report on her morning mishap. She wrote that someone on the graveyard shift had left her work area in disarray, including a misaligned load tray. She also indicated that the design of the cart handle made damage likely; in the past other cart handles had been broken. She recommended a re-design of the handle brackets and immediate discipline for the graveyard shift in her work area.

The fact that Judy filled out a near-hit report is certainly good news. But was this a complete report? Were there some person factors within Judy that could have influenced the incident? Was Judy under stress or distress, and if so, could this have been a contributing factor? Actually, it's been estimated that from 75 to 85 percent of all industrial injuries can be partially attributed to inappropriate reactions to stress. Furthermore, stress-related headaches are the leading cause of lost work time in the United States.

Judy's near-hit report was also clearly biased by common attribution errors researched by social psychologists, and used by all of us at some time to deflect potential criticism and reduce distress. Attribution errors, along with stress and distress, represent potential barriers to achieving a Total Safety Culture.

What is Stress?

In simple terms, stress is a psychological and physiological reaction to events or situations in our environment. Whatever triggers the reaction is called a stressor. So stress is the reaction of our mind and body to *stressors* such as demands, threats, conflicts, frustrations, over-loads, or changes.

Figure 6.1 depicts an office worker reacting to work over-load—an almost everyday stressor. "Lean and mean" organiza-

Figure 6.1. Technology cannot always substitute for personnel.

tions can create very stressful circumstances. The quick-fix solution to cutting costs by reducing headcount can lead to long-term stress and safety problems, unless employees can adjust to increased or shifted responsibilities.

Let's return to the basics of stress. When you interpret a situation as being stressful, your body prepares to deal with it. This is the fight-or-flight syndrome, a process controlled through our sympathetic nervous system. Adrenaline rushes into the bloodstream, the heart pumps faster, and breathing increases. Blood flows quickly from our abdomen to our muscles—causing those "butterflies in our stomach." We can also feel tense muscles or nervous strain in our back, neck, legs, and arms.

CONSTRUCTIVE OR DESTRUCTIVE?

The terms stress and distress commonly carry negative connotations. Something unwanted and bad is happening. But the first definition of stress in my *American Heritage Dictionary* (1992) is "importance, significance, or emphasis placed on something" (p. 1205). Similarly, the *New Merriam-Webster Dictionary* (1989) defines stress as "a factor that induces bodily or mental tension . . . a state induced by such a stress . . . urgency, emphasis" (p. 701). In contrast, distress is defined as "anxiety or suffering . . .severe strain resulting from exhaustion or an accident" (*The American Heritage Dictionary*, 1991, p. 410), or "suffering of body or mind; pain, anguish; trouble, misfortune...a condition of desperate need (*The New Merriam-Webster Dictionary*, 1989, p. 224).

Psychological research supports these distinctions between stress and distress. Stress can be positive, giving us heightened awareness, sharpened mental alertness, and an increased readiness to perform. Certain psychological theories presume that some stress is necessary for people to perform. The person who asserts, "I work best under pressure," understands the motivational power of stress. But can too much pressure, too many deadlines, be destructive?

I'm sure most of you have been in situations—or predicaments—where the pressure to perform seemed overwhelming. This is the point where too much pressure can hurt performance, where stress becomes *dis*tress. The relationship between external stimulation or pressure to perform and actual performance is depicted in Figure 6.2. This inverted U-shaped function is known as the Yerkes-Dodson Law (Yerkes & Dodson, 1908).

The Yerkes-Dodson law states that, up to a point, performance will increase as arousal, or pressure to perform well, increases. But the best performance comes when arousal is optimum rather than maximum. Push a person too far and performance starts to deteriorate. In fact, at exceptionally high levels of pressure or tension a person might perform as poorly as when they are hardly stimulated at all. Ask someone who is hysterical, and some-

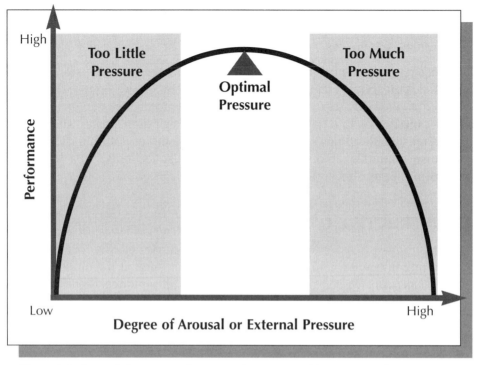

Figure 6.2. Arousal from external pressure (or a stressor) improves performance to an optimal point.

one who is about to fall asleep, to do the same job, and you won't be pleased by either result. Hans Selye, the Austrian-born founder of stress research said, "Complete freedom from stress is death" (Selye, 1974, p.32). It is extreme, disorganizing stress we need to avoid. Watch out for distress.

THE EYES OF THE BEHOLDER

Perceptions play an important role in stress and distress. The boss gives a group of employees a deadline; some tighten up inside, others take it in stride. Some circle their calendars and can't take their minds off the due date. Others seem to pay it no mind.

When a stressor is noticed and causes a reaction, the result can be constructive or destructive. If we believe we are *in control*—that we can deal with the overload, frustrations, conflicts, or whatever is triggered by the stressor—we become aroused and motivated to go beyond the call of duty. We actually achieve more. On the other hand, when we believe we can't handle the demands of the stressor, the resulting psychological and physiological reactions are likely to be detrimental to performance, and to our health and safety.

Why does the boss's deadline motivate one person and distress another? It depends on a number of internal "person factors." These include the amount of arousal already present in the individual and the person's degree of preparedness or self-confidence.

Those "butterflies" we feel in our stomach can help or hinder performance, depending on personal perception. When the "butterflies" are aligned for goal-directed behavior, we feel in control of the situation: prepared for it, challenged by it. The stress is positive—it arouses or motivates performance improvement. When the "butterflies" are misaligned and scattered in different directions, we feel unprepared, burdened rather than challenged. Stress is experienced as distress. This arousal can divert our attention, interfere with thinking processes, disrupt performance and reduce our ability and overall motiviation to perform well. The result can be at-risk behavior and a serious injury.

Identifying the Stressors

Stress or distress can be provoked by a wide range of demands and circumstances. Some stressors are acute, sudden life events, such as death or injury to a loved one, marriage, marital separation or divorce, birth of a baby, failure in school or at work, and a job promotion or relocation. Other stressors include the all-too-frequent minor hassles of everyday life, from long lines and excessive traffic to downsized work conditions and worries about personal finances.

Our jobs or careers are filled with stressors. Consider for a moment how much time you spend working, or thinking about your work. Some workplace stressors are obvious; others might not be as evident but are just as powerful. Work overload can obviously become a stressor and provoke either stress or distress, but what about "work underload"? Being asked to do too little can produce profound feelings of boredom, which can also lead to distress. Performance appraisals are stressors that can be motivational if perceived as objective and fair, or they can contribute to distress and inferior performance if viewed as subjective and unfair.

Other work-related factors that can be perceived as stressors and lead to distress and eventual burnout include: role conflict or ambiguity; uncertainty about one's job responsibilities; responsibility for others; a crowded, noisy, smelly, or dirty work environment; lack of involvement or participation in decision making; interpersonal conflict with other employees; and insufficient support from coworkers.

Let's review the key points about stress and distress. Actually, the flow schematic in Figure 6.3 says it all. First an environmental event is perceived and appraised as a stressor to be concerned about, or as a harmless or irrelevant stimulus. An event is perceived as a stressor if it involves *harm* or loss that has already occurred, a *threat* of some future danger, or a *challenge* to be overcome.

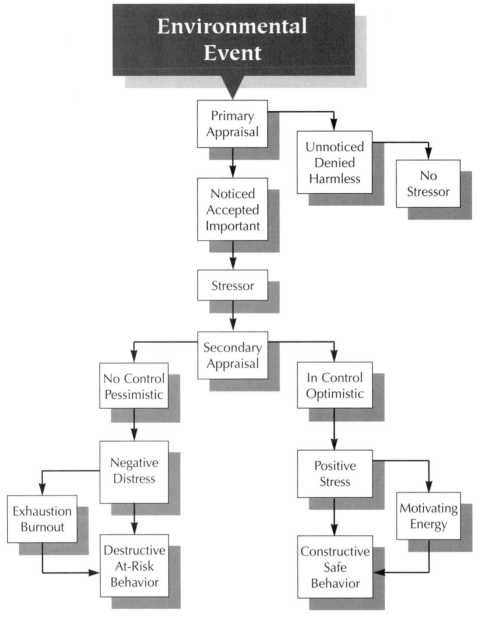

Figure 6.3. Through personal appraisal, people transform stressors into positive stress or negative distress.

Harm is how we appraise the impact of an event. For example, if you oversleep and miss an important safety meeting, the damage is done. In contrast, *threat* is how we assess potential future harm from the event. Missing the safety meeting could lower your team's opinion of you and reduce your opportunity to get actively involved in a new safety process. *Challenge* is our appraisal of how well we can eventually profit from the damage done. You could view missing the safety meeting as an opportunity to learn from one-on-one discussions with coworkers. This could demonstrate your personal commitment to the safety process and allow you to collect diverse opinions.

In this case, you are perceiving the stressor as an opportunity to learn and show commitment. This evaluation occurs during the secondary appraisal stage, and the result can be positive and constructive. On the other hand, your appraisal could be downbeat—you see no recourse for missing the safety meeting, and so you do nothing about it. Now coworkers might think you don't care about the new safety process and withdraw their support. In turn, you might give up or actively resist participating. This outcome would be nonproductive, of course, and possibly destructive.

The secondary appraisal stage, as depicted in Figure 6.3, determines whether the stressor leads to positive stress and constructive behavior, or to negative distress and destructive behavior. The difference rests with the individual. Does he or she assess the stressful situation as controllable, and thus remain optimistic during attempts to cope with the stressor? Or are the stressors seen as uncontrollable and unmanageable? If so, a helpless or pessimistic attitude can prevail and lead to distress and destructive or at-risk behavior. Several personal, interpersonal and environmental factors influence whether this secondary appraisal leads to constructive or destructive behavior. This is the theme of the next section.

Coping with Stressors

Understanding the multiple causes of conflict, frustration, overload, boredom, and other potential stressors in our lives can sometimes lead to effective coping mechanisms. These include:

- Revising schedules to avoid hassles like traffic and shopping lines
- Refusing a request that will overload us
- Finding time to truly relax and recuperate from tension and fatigue
- Communicating effectively with others to clarify work duties, reduce conflict, gain support, or feel more comfortable about added job duties
- Getting reassigned to a task that better fits our present talents and aspirations

The fact is, though, it's often impossible to avoid sudden (acute) or continuous (chronic) stressors in our lives. We need to deal with these head-on. Believing you can handle the harm, threat, or challenge of stressors is the first step toward experiencing stress rather than distress, and acting constructively rather than destructively.

PERSON FACTORS

Certain personality characteristics make some people more resistant to distress. Individuals who believe they control their own destiny and generally expect the best from life are, in fact, more likely to gain control of their stressors and experience positive stress rather than distress. It's important to realize that these person factors— self-mastery and optimism—are not permanent inborn traits. They are states of mind or expectations derived from personal experience, and they can be nurtured. It's possible to give people experiences that increase feelings of being "in control," experiences that lead people to believe something good will come from their attempts to turn stress into constructive action.

LEARNING TO FEEL HELPLESS. When I help clients assess the safety climate of their workplaces, I often uncover an attitude among hourly workers, and some managers as well, that reflects an important psychological concept called "learned helplessness." For instance, when I ask workers what they do regularly to make their workplace safer, I often hear:

"Besides following the safety procedures there's not much I can do for safety around here."

"It really doesn't matter much what I do, whatever will be will be."

"There's not much I can do about reducing work injuries; if it's my time, it's my time."

This is learned helplessness. The concept was labeled more than 20 years ago by research psychologists studying the learning process of dogs (Maier & Seligman, 1976; Seligman, 1975). They measured the speed at which dogs learned to jump a low barrier separating two chambers in order to avoid receiving an electric shock through the grid floor. A tone or light would be activated, then a shock was applied to the grid floor of the chamber. At first the dogs didn't jump the hurdle until receiving the shock, but after a few trials the dogs learned to avoid the shock by jumping into the other chamber as soon as the warning signal was presented.

Some dogs experienced shocks regardless of their behavior before the regular shock-avoidance learning trials. These dogs did not learn to jump the barrier to escape the shock. Instead, they typically just laid down in the shock chamber and whimpered. The earlier bad experience with inescapable shocks had taught the dogs to be helpless. Dr. Martin Seligman and associates coined the term "learned helplessness" to describe this state. Their finding has been demonstrated in a variety of human experiments as well.

Note how prior failures conditioned experimental subjects ranging from dogs to humans to feel helpless, in fact to be helpless. It's rather easy to assume that workers develop a "helpless" perspective regarding safety as a result of bad past experiences. If safety suggestions are ignored, or policies and procedures always come from management, workers might learn to feel helpless about safety. It's also true, however, that life experiences beyond the workplace can shape an attitude of learned helplessness. Certain individuals will come to work with a greater propensity to feel helpless in general, and this can carry over to feelings regarding occupational safety and health.

LEARNED OPTIMISM. A bad experience does not necessarily lead to an attitude of learned helplessness. You probably know people who seem to derive strength or energy from their failures, and try even harder to succeed when given another chance. Similarly, Dr. Seligman and colleagues found that certain dogs resisted learned helplessness if they previously had success avoiding the electric shock. So it is that some people tend to give up in the face of a stressor, while others fight back.

What can be done to help those who feel helpless? How can we get them to commit to and participate in the proactive processes of injury prevention? The work climate can play a critical role here. This happens when employees are empowered to make a difference and perceive they are successful.

When workers believe through personal experience that their efforts can make a difference in safety, they develop an antidote for learned helplessness. This has been termed "learned optimism." If the corporate climate empowers workers to take control and manage safety for themselves and their coworkers, they can legitimately attribute safety success to their own actions. This bolsters learned optimism and feelings of being in control.

FIT FOR STRESSORS

Fitness is another way to increase our sense of personal control and optimism. Being physically fit increases our body's ability to cope with the fight-or-flight syndrome discussed earlier. You probably know the basic guidelines for improving fitness, which include: stop smoking; reduce or eliminate alcohol consumption; exercise regularly, at least three times a week for about 30 minutes per session; eat balanced meals with decreased fat, salt, and sugar, and don't skip breakfast; and obtain enough sleep (usually 7 or 8 hours per 24 hour period for most people). Some of us find that following these guidelines over the long haul is easier said than done. We need support and encouragement to break a smoking or drinking habit, or to maintain a regular exercise routine.

Figure 6.4 illustrates the type of behavior that has come with the computer revolution. Low physical activity has become the way of work life for many of us. Often this inactivity spills over into home life. Survey research has shown that only one in five Americans exercises regularly and intensely enough to reduce the risk of stressor-induced heart disease. Figure 6.4 also depicts smoking behavior, considered to be the largest preventable cause of illness and premature death (before age 65) in the United States, accounting for approximately 125,000 deaths each year.

Also portrayed in Figure 6.4 is the positive influence of perceived control. Although the behavior is essentially the same at work (10 to 5) and at home (5 to 10), the individual is seemingly much happier at home. Why? Because at home he holds the remote control and therefore perceives more personal control.

But personal control is truly in the eyes of the beholder. Figure 6.5 depicts legitimate perceptions of control from the subjects of an experiment. These rodents are not usually considered "in control" of the situation, but in many ways they are. By simply changing our perspective, we can often perceive and accept more personal control at work. And this can turn negative distress into positive stress.

Figure 6.4. Becoming a "mouse potato" by day and a "couch potato" by night can reduce our physical ability to cope with stressors.

SOCIAL FACTORS

A support system of friends, family, and coworkers can do wonders in helping us reduce distress in our lives. Social support can motivate us to do what it takes to stay physically fit. And the people around us can make a boring task bearable and even satisfying. Of course they can also turn a stimulating job into something dull and tedious. It works both ways. People can motivate us, or trigger conflict, frustration, hostility, a win/lose perspective, and distress. It's up to us to make the most of the people around us.

Figure 6.5. Even the most obvious top-down situation allows for perceptions of bottom-up control.

We can learn from those who take effective control of stressful situations and expect the best. Or we can listen to the complaining, backstabbing, and cynicism of others, and fuel our own potential for distress.

It's obviously important to interact with those who can help us build resistance against distress and help us feel better about potential stressors. We can also set the right example, and be the kind of social support to others that we want for ourselves. The good feelings of personal control and optimism you experience from reaching out to help others can do wonders in helping you cope with your own stressors. And this actively caring stance builds your own support system, which you might need if your own stressors get too overwhelming to handle yourself.

The next section of this chapter introduces another means of reducing distress. It's a phenomenon that has particular implications for safety. In the aftermath of an injury or near hit, it can distort reports and investigations. The result: inappropriate or less-than-optimal suggestions for corrective action. This phenomenon of attributional bias can also create communication barriers between people, and limit the cooperative participation needed to achieve a Total Safety Culture.

Attributional Bias

Think back to the anecdote at the start of this chapter. I suggested that Judy's near-hit report was incomplete or biased. Specifically, Judy did not report the potential influence of her own distress on the incident. Rather, she focused on factors outside her immediate control—the poor bracket design for the wooden handle and the messy work area left by others. Giving up personal responsibility eliminated the incident as a stressor for her. She didn't have to deal with any guilt for almost hurting herself and damaging property. Her denial eased her distress, but biased the near-hit report. Psychologists refer to this as an attributional bias. By understanding when and how this phenomenon occurs, we can focus injury investigations on finding facts—not faults. This is a paradigm shift needed to achieve a Total Safety Culture.

THE FUNDAMENTAL ATTRIBUTION ERROR

Every day, we struggle to explain the actions of others. Why did she say that to me? Why did the job applicant refuse to answer that question? Why did Joe leave his work station in such a mess? Why did the secretary hang up on me? Why did Gayle take sick leave? Why does she allow her young children to ride in the bed of her pick-up truck? Why did the motorist pull a gun out of his glove compartment to shoot someone in the next car? Why were Nicole Brown-Simpson and Ronald Goldman murdered so brutally? In trying to answer questions like these, we point to external, environmental factors, such as equipment malfunctioning, excessive traffic, warm climate, and work demands; or to internal, person factors, such as personality, intelligence, attitude, or frustration.

Social psychologists have discovered a "fundamental attribution error" when systematically studying how people explain the behavior of others. When evaluating others, we tend to overestimate the influence of internal factors and underestimate external factors. We're more apt to judge the job applicant as rude or unaware (internal factors) than caught off-guard by a confusing or unclear question (external factor). Joe was sloppy or inconsiderate rather than overwhelmed by production demands. The injured employee was careless rather than distracted by a sudden environmental noise.

That's how we see things when we're judging others. It's different when we evaluate ourselves. The individuals performing the behaviors in the previous paragraph would say the causes were due more to external than internal factors.

Here's an example. My university students are quick to judge me as being an extrovert—outgoing and sociable. When I lecture in large classes of 600 to 800 students I'm animated and enthused, and they attribute my performance to internal personality traits. But I know better. I see myself in many different situations, and realize just how much my behavior changes depending on where I am. In many social settings I'm downright shy and reserved. I'm very sensitive to external influences.

THE SELF-SERVING BIAS

Students who flunk my university exams are quick to blame external factors, like tricky questions, wrong reading material assignments, and unfair grading. In contrast, students who do well are quite willing to give themselves most of the credit. It wasn't that I taught them well or that the exam questions were straightforward and fair; rather the student is intelligent, creative, motivated, and prepared. This real-world example, which I bet most readers can relate to, illustrates another type of attributional distortion, referred to as the self-serving bias.

How does this bias affect incident or injury investigation? Think of Judy's near-hit experience. She protected her self-esteem by overestimating external causes, and underplaying internal factors. This aspect of the self-serving bias is illustrated clearly and humorously by the list of explanations for vehicle crashes given in Figure 6.6. These external and situational excuses were taken from actual insurance forms submitted by the drivers.

People will obviously go to great lengths to shake blame for unintentional property damage or injury. This reduces negative stress or distress. No one wants to feel responsible for a workplace injury, especially if the company puts heavy emphasis on reducing "numbers," such as the plant's total recordable injury rate.

It's important for us to acknowledge how perceptions can be biased. Outsiders tend to blame the victim; victims look to extenuating circum-

- The other car collided with mine without giving warning of its intentions.
- A pedestrian hit me and went under my car.
- The guy was all over the road. I had to swerve a number of times before I hit him.
- I had been shopping for plants all day, and was on my way home. As I reached an intersection, a hedge sprang up, obscuring my vision. I did not see the other car.
- As I approached the intersection, a stop sign suddenly appeared in a place where no stop sign had ever appeared before. I was unable to stop in time to avoid the accident.
- An invisible car came out of nowhere, struck my vehicle, and vanished.
- My car was legally parked as it backed into the other vehicle.
- The pedestrian had no idea which direction to go, so I ran over him.
- The telephone pole was approaching fast. I was attempting to swerve out of its path when it struck my front end.

Figure 6.6. People are reluctant to admit personal blame for their vehicle crashes. (**Excerpted from the *Toronto Sun*, 1977**)

stances. We should empathize with the self-serving bias of the victim because it will reduce the person's distress. It will shift attention to external factors that can be controlled more easily than internal factors related to a person's attitude, mood, or state of mind.

In Conclusion

In this chapter I explained the difference between stress and distress, and discussed some strategies for reducing distress or turning negative distress into positive stress. Stress and distress begin with a stressor, which can be a major life event or a minor irritation of everyday living. You can evaluate or appraise the stressor in a way that is constructive, resulting in safe behavior; or destructive, causing at-risk behavior. When people are physically fit, in control, optimistic, and able to rely on the social support of others, they are most likely to turn a stressor into energy to achieve success. This is positive stress.

When stressors are perceived as insurmountable and unavoidable, distress is likely. Without adequate support from others, this condition can lead to physical and mental exhaustion, at-risk behavior, and unintentional injury to oneself or others. We need to become aware of the potential stressors in our lives and in the lives of our coworkers. In addition, we need to develop personal and interpersonal strategies to prevent distress in ourselves and others.

Victims of a near hit or injury will likely feel stressed during their primary appraisal. If their secondary appraisal clarifies the incident as an uncontrollable failure, negative stress or distress is likely. But they could interpret it as an opportunity to collect facts, learn, and implement an action plan to prevent a recurrence. Now the victim is experiencing positive stress, and constructive behavior is likely.

The work culture, including policies, paradigms, and personnel, can have a dramatic impact on whether victims of near hits, injuries, or other adversities experience stress or distress. The fundamental attribution error, where we overestimate personal factors to explain others' behavior ("Judy broke the handle because she was tired, stressed-out, and careless") can provoke distress, and pinpoint the very aspects of an incident most difficult to define and control. Instead, a victim's natural tendency to reveal a self-serving bias when discussing the incident—by putting more emphasis on external, situational causes—should be supported by the work culture. This reduces the victim's distress and puts the focus on the observable factors, including behavior, most readily defined and influenced. I detail processes for doing just that in Part Three.

Part Three

Behavior-Based Psychology

Chapter 7
Basic Principles

To achieve a Total Safety Culture, we need to integrate behavior-based and person-based psychology and affect large-scale culture changes. The five chapters in Part Three explain principles and procedures founded on behavior-based research which can be applied successfully to change behaviors and attitudes throughout organizations and communities. This chapter describes the primary characteristics of the behavior-based approach to the prevention and treatment of human problems, and shows their special relevance to occupational safety. The three basic ways we learn are reviewed, and related to the development of safe versus at-risk behaviors and attitudes.

Specific safety techniques can be viewed as possible routes to reach a destination, in our case a Total Safety Culture. A particular route may be irrelevant or need to be modified substantially for a given work culture. The key is to begin with a complete and accurate map. In other words, it's most important to start with an understanding of the basic principles.

If you recall, our overall map or guiding principle is represented by the Safety Triad (Figure 2.2). Its reference points are the three primary determinants of safety performance—environment, person, and behavior factors. To achieve a Total Safety Culture, we need to understand and pay attention to each.

In Part Two, I addressed a number of person-based factors that can contribute to injuries, including cognitions, perceptions, and attributions. The BASIC ID acronym was introduced in Chapter 4 to express the complexity of human dynamics and the special challenges involved in preventing injuries. Behavior was the first dimension discussed, and it is implicated directly or indirectly in each of the other dimensions. Attitudes, sensations, imagery, and cognitions—the thinking, person side of the Safety Triad—are each influenced by behavior. That's what is meant by the phrase, "Acting people into changing their thinking." When we change our behaviors, such as adopting a new strategy or procedure, certain person factors change too.

The reverse is also true. Changes in attitudes, sensations, imagery, and cognitions can alter behaviors. But considerable research has shown that it's easier and more cost-effective to "act people into changing their thinking" than the reverse, especially in organizations and community settings.

Primacy of Behavior

Whether *treating* clinical problems (such as drug abuse, sexual dysfunction, depression, anxiety, pain, hypertension, and child or spouse abuse) or *preventing* any number of health, social, or environmental ills (from developing healthy and safe lifestyles to improving education and protecting the environment), overt behavior is the focus. Treatment or prevention is based on three basic questions:

1. What behaviors need to be increased or decreased to treat or prevent the problem?

2. What environmental conditions, including interpersonal relationships, are currently supporting the undesirable behaviors or inhibiting desirable behaviors?

3. What environmental or social conditions can be changed to decrease undesirable behaviors and increase desirable behaviors?

Thus, behavior change is both the outcome and the means. It is the desired outcome of treatment or prevention, and the means to solving the identified problem.

REDUCING AT-RISK BEHAVIORS

Heinrich's well-known Law of Safety implicates at-risk behavior as the root cause of most near hits and injuries (Heinrich *et al,* 1980). Over the past 20 years, various behavior-based research studies have verified this aspect of Heinrich's Law by systematically evaluating the impact of interventions designed to lower employees' at-risk behaviors. Feedback from behavioral observations was a common ingredient in most of the successful intervention processes, whether the feedback was delivered verbally, graphically by tables and charts, or through corrective action.

The behavior-based approach to reducing injuries is depicted in Figure 7.1. At-risk behaviors are presumed to be a major cause of a series of progressively more serious incidents, from a near hit to a fatality. According to Heinrich's Law, there are numerous risky acts for every near hit, and many more near hits than lost-time injuries. This is good news. But let's not forget that timing or luck is usually the only difference between a near hit and a serious injury.

Typically, behavior change techniques are applied to specific targets. It's necessary, of course, that participants know why targeted behaviors are unde-

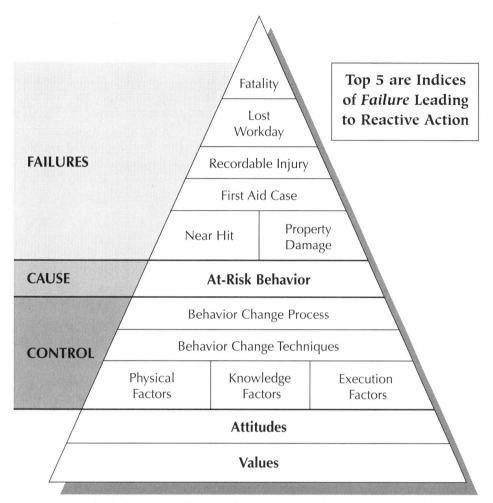

Figure 7.1. The behavior-based approach decreases at-risk behaviors to avoid failure.

sirable and have the physical ability to avoid them. Education and engineering interventions are sometimes needed to satisfy the physical and knowledge factors of Figure 7.1. The execution factors represent the motivational aspect of the problem, and usually require the most attention. In other words, people usually know what at-risk behaviors to avoid and have the ability to do so, but their motivation might be lacking or misdirected. Behavior change techniques are used to align individual and group motivation with avoiding the undesired at-risk behavior.

Values and attitudes form the foundation of the pyramid in Figure 7.1. These obviously critical person factors need to support the safety process. Remember our discussion about risk compensation in Chapter 5, and Professor Wilde's warning that it's more important to reduce risk tolerance than increase compliance with specific safety rules (Wilde, 1994)? This happens when people believe in the safety process and help to make it work.

Behavior helps to make the process work, and if involvement is voluntary and appropriately rewarded, it will lead to supportive attitudes and values to keep the process going.

The behavior-based approach illustrated in Figure 7.1 is failure oriented. It's also more reactive than proactive. The outcome measures are failures— fatalities, lost workdays, and the like—that require a fix. It's better to focus on increasing safe behaviors. This is being proactive; after all, behaviors are the root of the failures. Plus, by emphasizing safe behaviors, employees are more positive about the process and more willing to participate.

Increasing Safe Behaviors

Figure 7.2 illustrates a positive and proactive behavior-based model. I do not recommend this instead of the corrective action approach depicted in Figure 7.1. A complete behavior-based process should target both what's right and wrong about observed behavior. But again, more people will participate with a positive attitude and remain committed over time if there is more recognition of achievements than correction of failures.

MONITORING ACHIEVEMENT. The indices of achievement in Figure 7.2 are generally more difficult to record and track than those in Figure 7.1. Actually, the failure outcomes in Figure 7.1 are observed and recorded quite naturally. Except for near hits and first aid cases, the failures in Figure 7.1 have traditionally resulted in systematic investigation and formal reports. In contrast, the achievements in Figure 7.2 are somewhat difficult to define and record. In fact, it's impossible to obtain an objective record of the number of injuries prevented. A reasonable estimate of injuries prevented can be calculated, though, after you achieve a consistent decrease in injuries as a result of a proactive, behavior-based process.

It's possible to derive direct and objective definitions of the other success indices in Figure 7.2, and use these to estimate overall achievement. Involvement, for example, can be defined by recording participation in voluntary programs, and incidents of corrective action can be counted in a number of situations. You can chart the number of safety work orders turned in and completed, the number of safety audits completed and safety suggestions given, and the number of safety improvements occurring as a result of near-hit reports.

SAFETY SHARE. The *safety share* noted in Figure 7.2 is a simple behavior-focused process that reflects my emphasis on achievement. At the start of group meetings, the leader asks participants to report something they have done for safety during the past week or since the last meeting. Because the safety share is used to open all kinds of meetings, safety is given special status and integrated into the overall business agenda. My experience is that

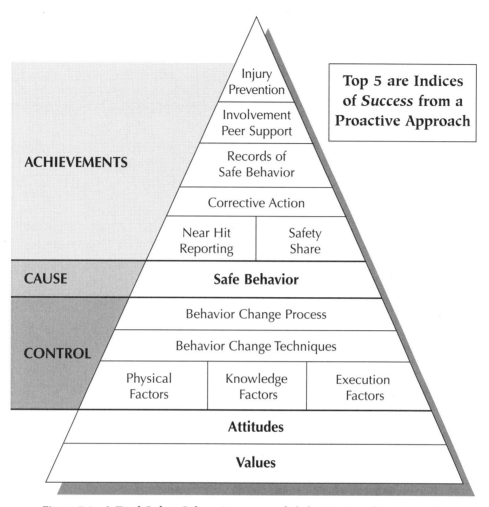

Figure 7.2. A Total Safety Culture increases safe behaviors to achieve success.

people come to expect queries about their safety accomplishments, and go out of their way to have an impressive safety story to share.

This simple awareness booster—"What have you done for safety?"—helps teach an important lesson. Employees learn that safety is not only loss control, an attempt to avoid failure, but can be discussed in the same terms of achievement as productivity, quality, and profits. As a measurement tool, it's possible to count and monitor the number of safety shares offered per meeting as an estimate of proactive safety success in the work culture.

Learning from Experience

A key assumption of behavior-based approach is that behavior (desirable and undesirable) is learned, and can be changed by providing people with new learning experiences. Diverse cultural, social, environmental, and bio-logical factors interact to influence our readiness to learn behaviors, and they also support behaviors once they are learned. We don't understand exactly how the particulars work—diverse factors interact to influence behaviors for each individual. The basic ways we develop behavioral patterns have been researched, though, and it's possible to identify principles behind learning and maintaining human behavior.

Psychologists define learning as a change in behavior, or potential to behave in a certain way, resulting from direct and indirect experience. In other words, we learn from observing and experiencing events and behaviors in our environment.

While the effects of learning are widespread and varied, it's generally believed that there are three basic models: *classical conditioning, operant conditioning,* and *observational learning.*

CLASSICAL CONDITIONING

This form of learning became the subject of careful study in the early twentieth century with the seminal research of Ivan Pavlov (1849-1936), a Nobel Prize-winning physiologist from Russia. Pavlov (1927) did not set out to study classical conditioning; rather, his research focused on the process of digestion in dogs. He was interested in how reflex responses were influenced by stimulating a dog's digestive system with food. During his investigations he serendipitously found that his subjects began to salivate before actually tasting the food. They appeared to anticipate the food stimulation by salivating when they saw the food or heard the experimenters preparing it. Some dogs even salivated when seeing the empty food pan or the person who brought in the food. Through experiencing the relationship between certain stimuli and food, the dogs learned when to anticipate food. Pavlov recognized this as an important phenomena and shifted the focus of his research to address it.

The top half of Figure 7.3 depicts the sequence of stimulus-response events occurring in classical conditioning. Actually, before learning occurs, the sequence includes only three events—conditioned stimulus (CS), unconditioned stimulus (UCS), and unconditioned response (UCR). The UCS elicits a UCR automatically, as in an autonomic reflex. That is, the food (UCS) elicited a salivation reflex (UCR) in Pavlov's dogs. In the same way, the smell of popcorn (UCS) might make your mouth water (UCR), a puff of air to your eye (UCS) would result in an automatic eyeblink (UCR), ingesting certain drugs (like Antabuse) would elicit a nausea reaction, and a state trooper writing you a speeding ticket is likely to influence an emotional reaction (like distress, nervousness, or anger).

If a particular stimulus (CS) consistently precedes the UCS on a number of occasions, the reflex (or involuntary response) will become elicited by the

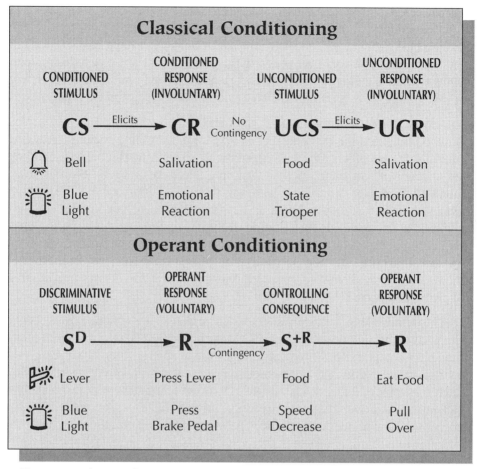

Figure 7.3. The stimulus-response relationships in classical and operant conditioning.

CS. This is classical conditioning, and occurred when Pavlov's "slobbering dogs" salivated when they heard the bell that preceded food delivery. Classical conditioning would also occur if your mouth watered when you heard the bell from the microwave oven tell you the popcorn is ready, if you blinked your eye following the illumination of a dim light that consistently preceded the air puff, if you felt nauseated after seeing and smelling the alcoholic beverage that previously accompanied the ingestion of Antabuse, and if you got nervous and upset after seeing a flashing blue light in your vehicle's rearview mirror.

OPERANT CONDITIONING

As shown in the lower portion of Figure 7.3, the flashing blue light on the police car might influence you to press your brake pedal and pull over. This is not an automatic reflex action, but is a voluntary behavior you'd perform in order to do what you consider appropriate at the time. In other words, you have learned (perhaps indirectly from watching or listening to others or from prior direct experience) to emit certain behaviors when you see a police car with a flashing blue light in your rearview mirror. Of course, you have also learned that certain driving behaviors (like pressing the brake pedal) will give a desired consequence (like slowing down the vehicle), and this will enable another behavior (like pulling over for an anticipated encounter with the police officer).

SELECTION BY CONSEQUENCES. B. F. Skinner (1904-1990), the Harvard professor who pioneered the behavior-based approach to solving societal problems, studied this type of learning by systematically observing the behaviors of rats and pigeons in an experimental chamber referred to as a "Skinner Box" (much to Professor Skinner's disappointment). Skinner termed the learned behaviors in this situation *operants* because they were not involuntary and reflexive, as in classical conditioning, but instead they *operated* on the environment to obtain a certain consequence. A key principle demonstrated in the operant learning studies is that voluntary behavior is strengthened (increased) or weakened (decreased) by consequences (events immediately following behaviors).

The relationship between a response and its consequence is a contingency, and this relationship explains our motivation for doing most everything we choose to do. Thus, the hungry rat in the Skinner Box presses the lever to receive food and the vehicle driver pushes the brake pedal to slow down the vehicle. Indeed, we all select various responses to perform daily—like eating, walking, reading, working, writing, talking, and recreating—to receive the immediate consequences they provide us. Sometimes we emit the behavior to achieve a pleasant consequence, such as a reward. Other times we perform a particular act to avoid an unpleasant consequence—a punisher or penalty. And we usually stop performing behaviors that are followed by penalties.

EMOTIONAL REACTIONS. As I discussed earlier in Chapter 3, we feel better when working for pleasant consequences than when working to avoid or escape unpleasant consequences. This can be explained by considering the classical conditioned emotions that naturally accompany the agents of reward versus penalty. How do you feel, for example, when a police officer flashes a blue light to signal you to pull over? When the instructor asks to see you after class? When you enter the emergency area of a hospital? When the dental assistant motions that "you're next"? When your boss leaves you a phone message to "see him immediately"? When you see your family at the airport after returning home from a long trip?

Your reactions to these situations depend, of course, on your past experiences. Police officers, teachers, doctors, dentists, and supervisors do not typically elicit negative emotional reactions in young children. However, through the association of certain cues with the consequences we experience in the situation, a negative emotional response or attitude can develop. And you don't have to experience these relationships directly. In fact, we undoubtedly learn more from observing and listening to others than from first-hand experience. This brings us to the third way in which we learn from experience—observational learning.

OBSERVATIONAL LEARNING

A large body of psychological research indicates that this form of learning is involved to some degree in almost everything we do. Whenever you do something a particular way because you saw someone else do it that way, or because someone showed you to do it that way, or because characters on television or in a video game did it that way, you are experiencing observational learning. Whenever you attribute "peer pressure" as the cause of someone taking up an unhealthy habit (like smoking cigarettes or drinking excessive amounts of alcohol) or practicing an at-risk behavior (like driving at excessive speeds or adjusting equipment without locking out the energy source), you are referring to observational learning. When you remind someone to set an example for others, you are alluding to the critical influence of observational learning.

VICARIOUS CONSEQUENCES. As children we learned numerous behavioral patterns by watching our parents, teachers, and peers. When we saw our siblings or schoolmates receive rewards like special attention for certain behaviors, we were more likely to copy that behavior, for instance. This process is termed vicarious, or indirect, reinforcement. At the same time, when we observed others getting punished for emitting certain behaviors, we learned to avoid these behaviors. This is referred to as vicarious punishment.

As adults we teach others by example. As illustrated in Figure 7.4, our children learn new

Where did you learn those bad words?

Figure 7.4. Children learn a lot from their parents through observational learning.

behavior patterns, including verbal behaviors, by watching us and listening to us. In this way they learn what is expected of them in various situations.

Our actions influence others to a greater extent than we realize. Without us being aware of our influence, children learn by watching us at home; and our coworkers are influenced by our practices at work. Not only does the performance of safe acts encourage similar behavior by observers, but verbal behavior can also be influential. If a supervisor is observed commending a worker for her safe behavior or reprimanding an employee for an at-risk practice, observ-

Figure 7.5. Intentionally and unintentionally, we teach others through our examples.

ers may increase their performance of similar safety behaviors (through vicarious reinforcement) or decrease the frequency of similar at-risk behavior (through vicarious punishment). Indeed, to make safe behavior the norm—rather than the exception—we must always set an example both in our own work practices and in the verbal consequences we offer coworkers following their safe and at-risk behaviors. Figure 7.5 offers a memorable pictorial regarding the influence of example-setting on observation learning.

The poem presented in Figure 7.6, "Setting Examples" by Forest H. Kirkpatrick, says it all. I recommend copying it and posting it for others to read. It's so easy to forget the dramatic influence we have on others by our own behaviors. Obviously, we need to take the slogan "walk the talk" very seriously. In fact, if we're not convinced a particular safeguard or protective behavior is necessary for us ("It's not going to happen to me"), we need to realize, at least, that our at-risk behaviors can endanger others.

For example, I never get in my vehicle believing a crash will happen to me, so my rationale for buckling my combination lap and shoulder belt is to set the right example for others, whether they're in the car with me or not. Understanding the powerful influence of observational learning, we should feel obligated to set the safe example whenever someone could see us.

The eye's a better teacher and more willing than the ear;
Fine counsel is confusing, but example's always clear;
And the best of all the preachers are the ones who live their creeds.
For to see the good in action is what everybody needs.

I can soon learn how to do it if you'll let me see it done;
I can watch your hands in action, but your tongue too fast may run;
And the lectures you deliver may be very wise and true.
But I'd rather get my lesson by watching what you do.

For I may not understand you and the high advice you give.
There's no misunderstanding how you act and how you live.

Figure 7.6. This poem, written by Forrest H. Kirkpatrick, illustrates the power of observational learning.

OVERLAPPING TYPES OF LEARNING

Laboratory methodologies have been able to study each type of learning separately, but the real world rarely offers such purity. In life, the usual situation includes simultaneous influence from more than one learning type. The operant learning situation, for example, is likely to include some classical (emotional) conditioning. As I indicated earlier, this is one reason rewarding consequences should be used more frequently than punishing consequences to motivate behavior change.

Remember, a rewarding situation (unconditioned stimulus) can elicit a positive emotional experience (unconditioned response), and a punitive situation (UCS) can elicit a negative emotional reaction (UCR). With sufficient pairing of rewarding or punishing consequences with environmental cues (such as a work setting, or particular people), the environmental setting (conditioned stimulus) can elicit a positive or negative emotional reaction or attitude (conditioned response). This can in turn facilitate (if it's positive) or inhibit (if it's negative) ongoing performance.

Figure 7.7 depicts a situation in which all three learning types occur at the same time. As discussed earlier and diagrammed in Figure 7.3, the blue flashing light of the police car signals drivers to press the brake pedal of their car and pull over. In this case, the blue light is considered a *discriminative stimulus* because it tells people when to respond in order to receive or avoid a consequence. Actually, drivers would apply their brakes to avoid punitive consequences, so this situation illustrates an avoidance contingency where drivers respond to avoid failure.

The flashing blue light might also serve as a conditioned stimulus eliciting a negative emotional reaction. This is an example of classical conditioning occurring simultaneously with operant learning. Our negative emotional reaction to the blue light might have been strengthened by prior observational learning. As a child, we might have seen one of our parents pulled over by a state trooper and subsequently observed a negative emotional reaction from our parent. The children in Figure 7.7 are not showing the same emotional reaction of the driver. Eventually, they will probably do so as a result of observational learning. Later, their direct experience as drivers will strengthen this negative emotional response to a flashing blue light on a police car.

Figure 7.7. Three types of learning occur in some situations.

In Conclusion

In this chapter I have reviewed the basic principles underlying a behavior-based approach to the prevention and treatment of human problems. The variety of successful applications of this approach was discussed, based on my personal experiences. The behavior-based principles—the primacy of behavior, direct assessment and evaluation, intervention by managers and peers, and three types of learning—were explained with particular reference to reducing personal injury.

Since at-risk behaviors contribute to most if not all injuries, a Total Safety Culture requires a decrease in at-risk behaviors. Organizations have attempted to do this by targeting at-risk acts, exclusive of safe acts, and using corrective feedback, reprimands, or disciplinary action to motivate behavior change. This approach is useful, but less proactive and less apt to be widely accepted than a behavior-based approach that emphasizes recognition of safe behaviors. It will be easier to get employees involved in safety achievement if credit is given for doing the right thing more often than reprimands for doing wrong.

The three types of learning are relevant for understanding safety-related behaviors and attitudes. Most of our safe and at-risk behaviors are learned operant behaviors, performed in particular settings to gain positive consequences or avoid negative consequences. Classical conditioning often occurs at the same time to link positive or negative emotional reactions with the stimulus cues surrounding the experience of receiving consequences. These cues include the people who deliver the rewards or penalties. We often learn what to do and what not to do by watching others receive recognition or correction for their operant behaviors. This is observational learning, an ongoing process that should motivate us to try to set the safe example at all times.

Chapter 8
Defining Critical Behaviors

The practical "how to" aspects of this book begin with this chapter. The overall process is called **DO IT***, each letter representing the four basic components of a behavior-based approach:* **D***efine target behaviors to influence;* **O***bserve these behaviors;* **I***ntervene to increase or decrease target behaviors; and* **T***est the impact of your intervention process. This chapter focuses on developing a critical behavior checklist for objective observing, intervening, and testing.*

Now the action begins. Up to this point I've been laying the groundwork (rationale and theory) for the intervention strategies described here and in the next three chapters. From this information you will learn how to develop action plans to increase safe behaviors, decrease at-risk behaviors, and achieve a Total Safety Culture.

Why did it take me so long to get here—to the implementation stage? Indeed, if you're looking for "quick-fix" tools to make a difference in safety you may have skipped or skimmed the first two parts of this text and started your careful reading here. I certainly appreciate that the pressures to get to the bottom line quickly are tremendous. But remember, there is no quick fix for safety. The behavior-based approach that is the heart of this book is the most efficient and effective route to achieving a Total Safety Culture. It is a never-ending continuous improvement process, one that requires ongoing and comprehensive involvement from the people protected by the process. In industry, these are the operators or line workers.

Long-term employee participation requires understanding and belief in the principles behind the process. Employees must also perceive that they "own" the procedures that make the process work. For this to happen it's necessary to teach the principles and rationale first (as done in this book), and then work with the process participants to develop specific procedures. This creates the perception of ownership and leads to long-term involvement.

As we begin here to define principles and guidelines for action plans, it's important to keep one thing in mind—you need to start with the conviction that there is rarely a generic best way to implement a process involving human interaction. For a behavior-based safety process to succeed in your setting, you will need to work out the procedural details with the people whose involvement is necessary. The process needs to be customized to fit your culture.

The DO IT Process

For the past several years I have taught applications of the behavior-based approach to industrial safety with the acronym: DO IT. The process is continuous and involves the following four steps:

D= Define critical target behavior(s) to increase or decrease

O= Observe the target behavior(s) during a preintervention baseline period to set behavior-change goals, and perhaps to understand natural environmental or social factors influencing the target behavior(s)

I = Intervene to change the target behavior(s) in desired directions

T = Test the impact of the intervention procedure by continuing to observe and record the target behavior(s) during the intervention program

From data recorded in the last item, you can evaluate the impact of your intervention and make an informed decision whether to continue it, implement another strategy, or define another behavior for the DO IT process.

This chapter focuses on the first two steps, defining and observing target behaviors. But before we get into those specifics, I want to briefly outline the DO IT process to make an important point: What I'm explaining to you is all easier said than done. Remember, there is no quick fix for safety.

To begin, just what are clear and concise definitions of target behaviors? This is the first step in the DO IT process. There is so much to choose from: Using equipment safely, lifting correctly, locking out power appropriately, and looking out for the safety of others, to name just a scant few. The outcome of behaviors, such as wearing personal protective equipment, working in a clean and organized environment, and using a vehicle safety belt can also be targeted. If two or more people independently obtain the same frequency recordings when observing the defined behavior or behavioral outcome during the same time period, you have a definition sufficient for an effective DO IT process. Baseline observations of the target behavior should be made and recorded before implementing an intervention program. More details on this aspect of the process are given later in this chapter.

What about the intervention step? This phase of DO IT involves one or more behavior-change techniques, based on the simple ABC model depicted

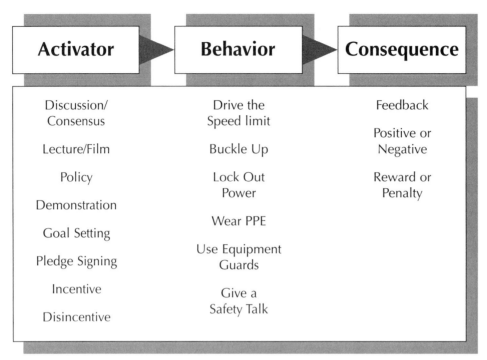

Activator	Behavior	Consequence
Discussion/ Consensus	Drive the Speed limit	Feedback
Lecture/Film	Buckle Up	Positive or Negative
Policy	Lock Out Power	Reward or Penalty
Demonstration		
	Wear PPE	
Goal Setting		
	Use Equipment Guards	
Pledge Signing		
Incentive	Give a Safety Talk	
Disincentive		

Figure 8.1. The ABC model is used to develop behavior change interventions.

in Figure 8.1. Activators direct behavior (as when a ringing telephone or doorbell signals the need for certain behaviors from residents), and consequences motivate behavior (residents answer or do not answer the telephone or door depending on current motives or expectations developed from prior experiences). Activators direct behavior and consequences motivate behavior.

Let's talk a little more about consequences. The strongest consequences are soon, sizable, and certain. In other words, we work diligently for immediate, probable, and large positive reinforcers or rewards; and we work frantically to escape or avoid soon, certain, and sizable penalties. This helps explain why safety is a struggle in many workplaces. You see, safe behaviors are usually not reinforced by soon, sizable, and certain consequences. In fact, safe behaviors are often punished by soon and certain negative consequences, including inconvenience, discomfort, and slower goal attainment. Also, the consequences that motivate safety professionals to promote safe work practices—reduced injuries and associated costs—are delayed, negative, and uncertain (actually improbable) from an individual perspective.

A few illustrations make these points. The man in Figure 8.2 will surely comply with the "Don't Walk" activator, and stay on the curb until the light changes. In this case, negative consequences from stepping off the curb at the wrong time would occur very soon. Given the observable traffic flow, a negative consequence is almost certain to happen, and the consequence could be quite sizable, perhaps fatal. But most safety situations in the workplace are not like this one.

Figure 8.2. Compliance with some activators is supported by natural consequences.

Figure 8.3. Compliance with some activators is not supported by natural consequences.

Take, for example, the man in Figure 8.3. He is complying with the activator, but there are no obvious (soon, certain, and sizable) consequences supporting this safe behavior. He might not see a reason for the hard hat, and without supportive consequences from peers his compliance could be temporary.

Competing with the lack of soon, certain, and sizable positive consequences for safe behaviors are soon, certain (and sometimes sizable) positive consequences for at-risk behaviors. Taking risks avoids the discomfort and inconvenience of most safe behaviors, and it often allows people to achieve their production and quality goals faster and easier. Supervisors sometimes activate and reward at-risk behaviors—unintentionally, of course—to achieve more production. Because activators and consequences are naturally available throughout our everyday existence to support at-risk behaviors in lieu of safe behaviors, safety can be considered a continuous fight with human nature (as discussed earlier in Part Two).

The DO IT process is a tool to use in this struggle with human nature. Developing and maintaining safe work practices often requires intervention strategies to keep people safe—strategies involving activators, consequences, or both. But we're getting ahead of ourselves. First we need to define critical behaviors as targets for our intervention. Let's see how this is done.

Defining Target Behaviors

The DO IT process begins by defining critical behaviors to work on. These become the targets of our intervention strategies. Some target behaviors might be safe behaviors you want to see happen more often, like lifting with knees bent, cleaning a work area, putting on personal protective equipment, or replacing safety guards on machinery. Other target behaviors may be at-risk behaviors that need to be decreased in frequency, such as misusing a tool, overriding a safety switch, placing obstacles in an area designated for traffic flow, stacking materials incorrectly, and so on. A DO IT process can define desirable behaviors to be encouraged or undesirable behaviors to be changed. What the process focuses on in your workplace depends on a review of your safety records, job hazard analyses, near-hit reports, audit findings, interviews with employees, and other useful information.

Determining which behaviors are critical is the first step of a DO IT process. A great deal can be discovered by examining the workplace and discussing with people how they have been performing their jobs. People already know a lot about the hazards of their work and the safe behaviors needed to avoid injury. They even know which safety policies are sometimes ignored to get the job done on time. They often know when a near hit occurred because an at-risk behavior or environmental hazard was overlooked. They also know which at-risk behaviors could lead to a serious injury (or fatality), and which safe behaviors could prevent a serious injury (or fatality).

In addition to employee discussions, injury records and near-hit reports can be consulted to discover critical behaviors (both safe and at risk). Job hazard analyses or standard operating procedures can also provide information relevant to selecting critical behaviors to target in a DO IT process. Obviously, the plant safety director or the person responsible for maintaining records for OSHA (Occupational Safety and Health Administration) can provide valuable assistance in selecting critical behaviors.

After selecting target behaviors, it's critical to define them in a way that gets everyone on the same page. All participants in the process must understand exactly what behaviors you intend to support, increase, or decrease. Defining target behaviors results in an objective standard for evaluating an intervention process.

WHAT IS BEHAVIOR?

The key is to define behaviors correctly. Let's begin by stepping back a minute to consider: What is behavior? Behavior refers to *acts* or *actions* by individuals that can be *observed by others*. In other words, behavior is what a person does or says, as opposed to what he or she thinks, feels, or believes.

Yes, the act of saying words such as "I am tired" is a behavior because it can be observed or heard by others. However, this is not an observation of tired behavior. If the person's work activity slows down or amount of time on the job decreases, we might infer that the person is actually tired. On the other hand, a behavioral "slow down" could result from other internal causes, like worker apathy or lack of interest. The important point here is that feelings, attitudes, or motives should not be confused with behavior. They are *internal* aspects of the person that cannot be directly observed by others. It is risky to infer inner personal characteristics from external behaviors.

OBSERVABLE BEHAVIORS

A target behavior needs to be defined in observable terms so that multiple observers can watch one individual independently and obtain the same results regarding the occurrence or nonoccurrence of the target behavior. There should be no room for interpretation. "Isn't paying attention," "acting careless," or "lifting safely," for example, are not adequate descriptions of behavior, because observers would not agree consistently about whether the behavior occurred. In contrast, descriptions like "keeping hand on handrail," "moving knife away from body when cutting," and "using knees while lifting" are objective and specific enough to obtain reliable information from trained observers. In other words, if two observers watched for the occurrence of these behaviors, they would likely agree whether or not the behavior occurred.

MULTIPLE BEHAVIORS

Let's look more closely at types of behaviors. Some workplace activities can be treated effectively as a single behavior. Examples include: "Looking left-right-left before crossing the road;" "Walking within the yellow safety lines;" "Honking the fork lift horn at the intersection;" "Returning tools to their proper location."

Some outcomes of behaviors also can be dealt with in singular terms: "Using ear plugs;" "Using a vehicle safety belt;" "Climbing a ladder that is properly tied off;" "Working on a scaffold with appropriate fall protection;" and "Repairing equipment that was locked-out correctly." With a proper definition, an observer could readily count occurrences of these safe behaviors (or outcomes) during a systematic audit.

But many safety activities are made up of more than one discrete behavior, and it may be important to treat these behaviors independently in a definition and an audit. "Bending knees while lifting," for example, is only one aspect of a safe lift. Thus, if safe lifting were the activity targeted in a DO IT process, it would be necessary to define the separate behaviors (or procedural steps) of a safe lift. This would include, at least, checking the load before lifting, asking for help in certain situations, lifting with the legs, holding the load close to one's body, lifting in a smooth motion, and moving feet when rotating (not twisting).

Each of the procedural steps in safe lifting requires a clear objective definition so two observers can determine reliably whether the behavior in a lifting sequence occurred. Observing reliably whether the load was held close and knees were bent would be relatively easy. However, defining a "smooth lift" so that observers could agree on 80 percent or more of the observations would be more difficult. And, for observers to reliably audit "asking for help," the "certain situations" calling for this response would need to be specified.

Observing Behavior

The acronym **SOON** depicted in Figure 8.4 reviews the key aspects of developing adequate definitions of critical behaviors to target for a DO IT process. We have already considered most of the characteristics of behavioral definitions implied by these key words, and examples of behavioral checklists are provided later in this chapter, as well as in Chapter 11 on "safety coaching."

Several methods are available for objectively observing and recording safe versus at-risk behaviors in real-world settings. Different situations call for different procedures. I'll give a number of examples throughout this and sub-

S PECIFIC
- Concise behavioral definition
- Unambiguous

O BSERVABLE
- Overt behaviors
- Countable & recordable

O BJECTIVE
- No interpretations nor attributions
- "What," not "Why"

N ATURALISTIC
- Normal interaction
- Real-world activity

Figure 8.4. Behavioral observations for the DO IT process should be "SOON."

sequent chapters. I hope at least one of these methods will relate directly to the situation in which you want to apply the DO IT process.

A PERSONAL EXAMPLE

My daughter Krista asked me to drive her to the local Virginia Department of Motor Vehicles office to get her "learner's permit." She was 15 years old and thought she was ready to drive. Of course, I knew better, but how do you fight a culture that puts teenagers behind the wheel of motor vehicles before they are really ready for such a risky situation?

"Don't worry, Dad," my daughter said, "I've had driver's education in high school." Actually, that was part of my worry. She was educated about the concepts and rules regarding driving, but she hadn't been trained. She had not yet translated her education into operations or action plans.

In order to obtain a license to operate a motor vehicle in Virginia before the age of 18, teenagers with a learner's permit are required to take seven two-hour instructional periods of on-the-road experience with an approved driver-training school. For half of these sessions they must be the driver; the rest of the time they sit in the back seat and perhaps learn through observation. Thus, for seven one-hour sessions Krista drove around town with an instructor in the front seat and one or more students in the back, waiting for their turn at the wheel. This was an opportunity for my daughter to transfer her driver education knowledge to actual performance.

DRIVING ACTIVATORS AND CONSEQUENCES. On-the-job training obviously requires an appropriate mix of observation and feedback from an instructor. Practice does not make perfect. Only through appropriate feedback can people improve their performance. Some tasks give natural feedback to shape our behavior. When we turn a steering wheel in a particular direction we see immediately the consequence of our action, and our steering behavior is naturally shaped. The same is true of several other behaviors involved in driving a motor vehicle, from turning on lights, windshield wipers, and cruise-control switches to pushing gas and brake pedals.

However, many other aspects of driving are not followed naturally with feedback consequences, particularly those that can prevent injury from vehicle crashes. Although we get feedback to tell us our steering wheel, gas pedal, turn signal lever, and brakes work, we don't get natural feedback regarding our safe versus at-risk use of such control devices. Also, when we first learn to drive we don't readily recognize the activators (or discriminative stimuli) that should signal the use of various vehicle controls. This is commonly referred to as "judgment." From a behavior-based perspective, "good driving judgment" is recognizing environmental conditions (or activators) that signal certain vehicle-control behaviors, and then implementing the controls appropriately.

I wondered whether my daughter's driving instructor would give her appropriate and systematic feedback regarding her "driving judgment." Would he point out consistently the activators that require safe vehicle-control behaviors? Would he put emphasis on the positive, supporting my daughter's safe behaviors before criticizing her at-risk behaviors? Would he, or a student in the back seat, display negative emotional reactions in certain situations and teach Krista (through classical conditioning) to feel anxious or fearful in particular driving situations? Would some at-risk driving behaviors by Krista or the other student drivers be overlooked by the instructor, and lead to observational learning that some at-risk driving behaviors are acceptable?

DEVELOPING A DRIVING BEHAVIOR CHECKLIST. Even if the driving instruction is optimal, seven hours of such observation and feedback is certainly not sufficient to teach safe driving habits. I recognized a need for additional driving instruction for my daughter. We needed a DO IT process for driving. The first step was to define critical behaviors to target for observation and feedback. Through one-on-one discussion, my daughter and I derived a list of critical driving behaviors and then agreed on specific definitions for each item. My university students practiced using this critical behavior checklist (CBC) a few times with various drivers and refined the list and definitions as a result. The CBC we eventually used is depicted in Figure 8.5.

After refining the CBC and discussing the final behavioral definitions with Krista, I felt ready to implement the second stage of DO IT: **O**bservation. I asked my daughter to drive me to the university—about nine miles from home—to pick up some papers. I made it clear that I would be using the CBC on both parts of the roundtrip. When we returned home I totaled the safe and at-risk checkmarks and calculated the percentage of safe behaviors. Krista was quite anxious to learn the results, and I looked forward to giving her objective behavioral feedback. I had good news. Her percentage of safe driving behaviors (percent safe) was 85 percent, and I considered this quite good for our first time.

I told Krista her percent safe score, and proceeded to show her the list of safe checkmarks (covering up the checks in the At-Risk column). Obviously, I wanted to make this a positive experience, and to do this, it was necessary to emphasize the behaviors I saw her do correctly. To my surprise, she did not seem impressed with her 85 percent safe score, and pushed me to tell her what she did wrong. "Get to the bottom line, Dad," she asserted, "Where did I screw up?" I continued an attempt to make the experience positive, saying "You did great, honey, look at the high number of safe behaviors." "But why wasn't my score 100 percent?" reacted Krista. "Where did I go wrong?"

This initial experience with the driving CBC was enlightening in two respects. It illustrated the unfortunate reality that the "bottom line" for many people is "where did I make a mistake?" My daughter, at age 15, had already

Critical Behavior Checklist for Driving

Driver:	Date:	Day:
Observer 1:	Origin:	Start Time:
Observer 2:	Destination:	End Time:
Weather:		
Road Conditions:		

Behavior	Safe	At-Risk	Comments
Safety Belt Use			
Turn Signal Use			
Left turn			
Right turn			
Lane change			
Intersection Stop			
Stop sign			
Red light			
Yellow light			
No activator			
Speed Limits			
25 mph and under			
25 mph-35 mph			
35 mph-45 mph			
45 mph-55 mph			
55 mph-65 mph			
Passing			
Lane use			
Following Distance (2 sec)			
Totals			

% Safe = $\dfrac{\text{Total Safe Observations}}{\text{Total Safe + At-Risk Observations}}$ = _____

Figure 8.5. A critical behavior checklist (CBC) to improve driving safety.

learned that people evaluating her performance seem to be more interested in mistakes than successes. That obviously makes performance evaluation (or appraisal) an unpleasant experience for many people.

A second important outcome from this initial CBC experience was the realization that people can be unaware of their at-risk behaviors, and only through objective feedback can this be changed. My daughter did not readily accept my corrective feedback regarding her four at-risk behaviors. In fact, she vehemently denied that she did not always come to a complete stop. However, she was soon convinced of her error when I showed her my data sheet and my comment regarding the particular intersection where there was no traffic and she made only a rolling stop before turning right. I did remind her that she did use her turn signal at this and every intersection, and this was something to be proud of. She was developing an important safe habit, one often neglected by many drivers.

I really didn't appreciate the two lessons from this first application of the driving CBC until my daughter monitored *my* driving. That's right, Krista used the CBC in Figure 8.5 to evaluate my driving on several occasions. I found this reciprocal application of a CBC to be most useful in developing mutual trust and understanding between us. I found myself asking my daughter to explain my lower-than-perfect score, and arguing about one of the recorded "at-risk" behaviors. I, too, was defensive about being 100 percent safe. After all, I've been driving for 37 years and teaching and researching safety for more than 20 years, how could I not get a perfect driving score when I knew I was being observed?

From our experience with the CBC, my daughter and I learned the true value of an observation and feedback process. While using the checklist does translate education into training through systematic observation and feedback, the real value of the process is the active caring that occurs between people. In other words, we learned not to get too hung up on the actual numbers. After all, there is plenty of room for error in the numerical scores. Rather we learned to appreciate the fact that through this process people are caring for the safety and health of each other in a way that can truly make a difference. We also learned that even experienced people can perform at-risk behavior and not even realize it.

FROM UNCONSCIOUS TO CONSCIOUS. Figure 8.6 depicts the process we often go through when developing safe habits. Both Krista and I were unaware of some of our at-risk driving behaviors. For these behaviors we were "unconsciously incompetent." Through the CBC feedback process however, we became aware of our at-risk driving behaviors. But awareness did not necessarily result in 100 percent safe behavior scores. Several feedback sessions were needed before some safe driving behaviors occurred regularly, and before some at-risk behaviors decreased markedly or extinguished completely. In other words, initial feedback made us "consciously incompetent" with regard to some driving behaviors. Continuous feedback and mutual support resulted in beneficial learning, as reflected in improved percent safe behavior scores on the CBC. Thus, for some driving behaviors we became "consciously competent."

Feedback made us aware of certain driving rules or the driving situation (activator) that calls for a particular safe behavior. This stage involves think-

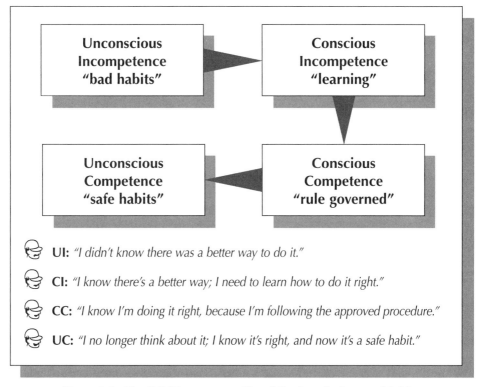

Figure 8.6. The DO IT process enables shifts from bad to good habits.

ing or talking to oneself to identify activators that require certain safe behaviors, and giving self-approval or self-feedback after performing the appropriate safe behavior. With continuous observation and feedback from both others and ourselves some safe behaviors become automatic or habitual. They reach the "unconscious competence" stage in Figure 8.6.

At what stage of habit formation are you when you get in the back seat of someone else's vehicle, like a taxi cab? It's possible to be "unconsciously competent" in some situations, but be "consciously competent" or "consciously incompetent" in another situation for the same behavior. For example, wearing safety glasses, ear plugs, and steel-toed shoes might be a safe habit on the job, but what are the safe behaviors when mowing the lawn in your backyard? A DO IT process could increase our use of personal protective equipment at home as well as at work. First, we need to accept the fact that we can all be unconsciously or consciously incompetent with regard to some behaviors. Then we need to understand the necessity of behavioral feedback to improve our performance. Thirdly, we need others in our family or work team to observe us with a CBC and then share their findings as actively caring feedback.

Two Basic Approaches

The CBC examples described above illustrate two basic ways of implementing the Define and Observe stages of DO IT. The driving CBC I developed with my daughter illustrates the observation and feedback process recommended by a number of successful behavior-based safety consultants. I refer to this approach as one-to-one safety coaching because it involves an observer using a CBC to provide instructive behavioral feedback to another person. This approach is reviewed briefly below; a more comprehensive description can be found in Chapter 11.

The second approach to the Define and Observe stages of DO IT involves a limited CBC (perhaps targeting only one behavior) and does not necessarily involve one-to-one coaching. This is the approach used in most of the published studies of the behavior-based approach to safety.

Each of these approaches to the Define and Observe stages of DO IT are advantageous for different applications within the same culture. Thus, it's important to understand the basic procedures of each, and to consider their advantages and disadvantages. For some work settings, I have found it quite useful to start with the simpler approach of targeting only a few CBC behaviors. With immediate success, behaviors are then added until eventually a comprehensive CBC is developed, accepted, and used willingly throughout the culture.

STARTING SMALL

This approach targets a limited number of critical behaviors, but does not require one-on-one observation. A work group defines a critical behavior or behavioral outcome to observe, as discussed earlier in this chapter. After defining their target so that two or more observers can reliably observe and record a particular property of the behavior, usually frequency of occurrence, the group members should give each other permission to observe this work practice among themselves. If some group members do not give permission, it's best not to argue with them. Simply exclude these individuals from the observations, and invite them to join the process whenever they feel ready. They will likely participate eventually when they see that the DO IT process is not a "gotcha program" but an objective and effective way to care actively for the safety of others and build a Total Safety Culture.

It often helps to develop a behavior checklist to use during observations. As discussed earlier, target behaviors like "safe lifting" and "safe use of stairs" include a few specific behaviors, either safe or at risk. Therefore, the CBC should list each behavior separately, and include columns for checking "safe" or "at risk." Figure 8.7 depicts a sample CBC for safe lifting. Through use of this CBC, a work group might revise the definitions, and possibly add a lifting-related behavior relevant to their work area.

Observer:		Date:	
TARGET BEHAVIOR		SAFE	AT RISK
load appropriate			
hold close			
use legs			
move feet; don't twist			
smooth motion; no jerks			

Comments (use back if necessary):

_____

_____

_____

_____

Figure 8.7. A sample behavior checklist to observe and record "safe lifting."

Participants willing to be observed anonymously for the target behavior(s) use the CBC to maintain daily records of the safe and at-risk behaviors defined by the group. They do not approach another individual specifically to observe them. Rather they look for opportunities for the target behavior to occur. When they see a safe behavior opportunity (SBO), they take out their checklist and complete it. If the target behavior is "safe lifting," for example, observers keep on the lookout for an SBO for lifting. They might observe such an SBO from their work station or while walking through the plant. Of course, if they see an at-risk lifting behavior and are close enough to reduce the risk, they should put their CBC aside and intervene. Intervening to reduce risk must take precedence over recording an observation of at-risk behavior.

OBSERVING MULTIPLE BEHAVIORS

As the list of targets on a CBC increases, it becomes more and more difficult to complete a checklist from a remote location. Auditing several critical behaviors usually puts observers in close contact with another person (the performer), resulting in a one-on-one coaching situation. The observer should seek permission from the performer before recording any observations, even though a work group might have agreed on the observation process in earlier education and training meetings. If the performer wishes not to be observed, the observer should leave with no argument and a friendly smile. This helps to build the trust needed to eventually reach 100 percent participation in the DO IT process.

Critical Behavior Checklist
for Personal Protective Equipment

Observation period (dates): _____

Observer: _____

TOTAL NUMBER OF EMPLOYEES OBSERVED	NUMBER OF EMPLOYEES OBSERVED USING ALL REQUIRED PPE

PPE (FOR OBSERVED AREA)	SAFE OBSERVATION (PROPER USE OF PPE)	AT-RISK OBSERVATION (IMPROPER OR NO USE OF PPE)
Gloves		
Safety Glasses/Shield		
Safety Shoes		
Hard Hat		
Lifting Belt		
Hearing Protection		
TOTAL		

Figure 8.8. A generic critical behavior checklist (CBC) for monitoring the use of personal protective equipment (PPE).

Multiple behavior CBCs might be specific to a particular job or be generic in nature. The driving CBC I used with Krista was a job-specific checklist, only relevant for operating a motor vehicle. In contrast, a generic checklist is used to observe behaviors that may occur at various job sites. The CBC depicted in Figure 8.8 is generic because it is applicable for any job that requires the use of personal protective equipment (PPE). Since different PPE might be required on different jobs, certain PPE categories on the CBC may be irrelevant for some observations. For jobs requiring extra PPE, additional behaviors will be targeted on the CBC. Obviously, the observer needs to know PPE requirements before attempting to use a CBC like the one shown in Figure 8.8.

The CBC in Figure 8.8 includes a place for the observer's name, but the performer's name is not recorded. Also, this CBC was designed to conduct several one-on-one behavior audits over a period of time. Each time the observer performs an observation he or she places a checkmark in the left box (for total number of observations). If the performer was using all PPE required in the work area, a check would be placed in the right-hand box. From these entrees, the overall percentage of safe employees can be monitored.

The checkmarks in the individual behavior categories of the CBC in Figure 8.8 are totaled, and by dividing the total number of safe checks by the total safe and at-risk checks, the percentage of safe behaviors for each PPE category can be assessed. This enables valuable feedback regarding the relative use of various devices to protect employees. Such information might suggest a need to make certain PPE more comfortable or convenient to use. It might also suggest the need for special intervention, as discussed in the next three chapters. Chapter 11 also includes additional information on the design of CBCs for one-on-one behavior observation.

In Conclusion

In this chapter we have gotten into the "nuts and bolts" of implementing a behavior-based safety process to develop a Total Safety Culture. The overall process is referred to as DO IT, each letter representing one of the four stages of behavior-based safety. This chapter focused on the first two stages: Define and Observe.

Defining critical behaviors to target for observation and intervention is not easy. A work team needs to consult a variety of sources, including the workers themselves, near-hit reports, injury records, job hazard analyses, and the plant safety director.

After selecting a list of behaviors critical to preventing injuries in their work area, the team needs to struggle through defining these behaviors so precisely that all observers agree on a particular property of each behavior at least 80 percent of the time. The behavioral property most often observed for industrial safety is frequency of occurrence, including a sampling of whether or not the target behavior occurred.

A critical behavior checklist (CBC) is used to observe and record the relative frequency (or percentage of opportunities) with which critical behaviors occur throughout a work setting. If the CBC contains only a few behaviors or behavioral outcomes (conditions resulting from behavior), it's possible to conduct observations without engaging in a one-on-one coaching session. This is often the best approach to use when first introducing behavior-based safety to a work culture. It's not as overwhelming or time-consuming as one-on-one coaching with a comprehensive CBC. Over time and through building trust, a short CBC can be readily expanded and lead to one-on-one safety coaching. Safety coaching is one very effective way to implement each stage of the DO IT process, and is detailed in Chapter 11. First, it's important to understand more about the third DO IT stage: Intervention. This is the topic of the next two chapters.

Chapter 9
Intervening with Activators

Intervention techniques to increase safe behaviors or decrease at-risk behaviors are either activators *or* consequences. *This chapter explains activators, with real-world examples showing how to develop effective strategies. Six principles for maximizing the impact of activators frame the discussion.*

Chapter 8 introduced the Activator-Behavior-Consequence (ABC) model as a framework for designing behavior-change interventions. Psychologists who use the behavior-based approach to solve human problems design activators (conditions or events preceding operant behavior) and consequences (conditions or events following operant behavior) to increase the probability that desired behaviors will occur and undesired behaviors will not. Activators precede and direct behavior. Consequences follow and motivate behavior. This chapter explains basic principles about activators to help you design interventions for increasing safe behavior and decreasing at-risk behavior. The next chapter focuses on the use of consequences to motivate safety achievement.

First, let me reiterate the need for safety interventions. As I've said before, maintaining our own safe behavior is not easy. It's usually one long fight with human nature, because in most situations activators and consequences naturally support risky behavior in lieu of safe behavior. At-risk behavior often allows for more immediate fun, comfort, and convenience than safe behavior, prompting the need for special intervention to direct and motivate safe behavior.

Activators are generally much easier and less expensive to use than consequences, so it's not suprising that they are employed much more often to promote safe behavior. Posters or signs are perhaps the most popular activators for safety:

Some bear only a general message—"Safety is a Condition of Employment"; others refer to a specific behavior—"Hard Hat Required in this Area"

Some signs request the occurrence of a behavior—"Walk," "Wear Ear Plugs in This Area"; others want you to avoid a certain behavior— "Don't Walk," "No Smoking Area"

Sometimes a relatively convenient response is requested—"Buckle Up," while other signs prompt relatively inconvenient behaviors—"Lock Out All Energy Sources Before Repairing Equipment"

Some signs imply consequences—"Use Eye Protection: Don't Be Blinded by the Light"; others do not—"Wear Safety Goggles"

We might be reminded of a general purpose—"Actively Care for a Total Safety Culture"; or challenged—"100 Percent Safe Behavior is Our Goal This Year"

I've visited a number of work environments where all of these types of safety signs were displayed. In fact, I've seen situations that make the illustration in Figure 9.1 seem not very far fetched. Does this sort of overkill work to change behavior and reduce injuries? If you answered "no," then this time your common sense was correct, because you've been there and experienced the ineffectiveness of many safety signs.

Which signs would you eliminate from Figure 9.1? How would you change certain signs to increase their impact? What activator strategies would you use instead of the signs? This chapter will enable you to answer these questions—not on the basis of common sense, but from behavioral science research. Let's move on to discuss six key principles for increasing the impact of activators.

Figure 9.1. Safety activators can be overwhelming and ineffective.

Principle #1: Specify Behavior

Behavioral research demonstrates that signs with general messages and no specification of a desired behavior to perform (or an undesirable behavior

Figure 9.2. Some activators are not specific enough.

Figure 9.3. Some signs are too complex to be effective.

to avoid) have very little impact on actual behavior. But signs that refer to a specific behavior can be beneficial.

Figure 9.2 illustrates "explosively" the need to include sufficient response information with a behavioral request. But too much specificity can bury a message, as illustrated in Figure 9.3. Activators ought to specify a desired response, but not overwhelm with complexity, as I've seen in a number of industrial signs. Overly complex signs are easy to overlook—with time, they just blend into the woodwork. Keeping signs salient or noticeable is clearly a challenge.

Principle #2: Maintain Salience with Novelty

HABITUATION

It is perfectly natural for activators like sign messages to lose their impact over time. This process is called *habituation,* and it's considered by some psychologists to be the simplest form of learning. Through habituation we learn not to respond to an event that occurs repeatedly. Habituation happens even among organisms with primitive nervous systems. For example, when you lightly tap the shell of a large snail it withdraws into its shell. After about 30 seconds the snail will extend its body from the shell and continue on its way. When you tap the shell again, the snail will withdraw again. However, this time the snail will stay inside its shell for a shorter duration. Your third tap will cause withdrawal again, but the withdrawal time will be even shorter. Each tap on the snail's shell results in successively shorter withdrawal time; eventually the snail will stop responding at all to your tap. The snail's behavior of withdrawal to the activator—shell tapping—will have habituated.

Habituation is perfectly consistent with an evolutionary perspective. If there is no obvious consequence (good or bad) from responding to a stimulus, the organism, be it an employee or a snail, stops reacting to it. It is a waste of time and energy to continue responding to an activator that seems to be insignificant. What would a snail do in a rain storm if it did not learn to ignore shell taps that have no consequence?

Consider the distractions and distress you'd experience daily if you could not learn to ignore noises from voices, radios, traffic, and machinery. At first these environmental sounds might be quite noticeable and perhaps distracting, but through habituation they become insignificant background noise. They no longer divert attention nor interfere with ongoing performance.

What is the relevance of habituation for safety? It's human nature to habituate to everyday activators in our environment that are not supported by consequences. And this is the case with most safety activators. Staying attentive to safety activators is a continuous fight with one aspect of human nature—habituation.

WARNING BEEPERS: A COMMON WORK EXAMPLE

Figure 9.4 illustrates quite clearly the phenomenon of habituation and reduced activator salience with experience. I'll bet you can reflect on personal experiences quite similar to the one shown here.

Not only has the brick mason habituated to the familiar "beep" of the backing vehicle, but the driver is illustrating danger compensation (or risk homeostasis) as I discussed in Chapter 6. He is not looking over his shoulder to check for a potential collision victim. He assumes the warning beeper is sufficient to activate coworkers' avoidance behavior and prevent injury. So this particu-

Figure 9.4. Some signals we rely on lose impact over time.

lar activator has actually reduced the driver's perceived risk. This influences his at-risk behavior of looking forward instead of turning his head to check his blindspots. A key point: Understanding the basic learning phenomenon of habituation can prevent over-reliance on activators and support a need to work more defensively.

Principle #3: Vary the Message

What does habituation tell us about the design of safety activators? Essentially, we need to vary the message. When an activator changes it can become more salient and noticeable. The "safety share" discussed in Chapter 7 follows this principle. When participants in a group meeting are asked to share something they've done for safety since the last meeting, the examples will vary considerably. Similarly, group discussions of near hits and potential corrective actions will also vary dramatically. The messages from safety shares and near-hit discussions will also be salient because they are personal, genuine, and distinct.

CHANGEABLE SIGNS

Over the years I've noticed a variety of techniques for changing the message on safety signs. There are removable slats to place different messages. I'm sure most of you have seen computer-generated signs with an infinite variety of safety messages. Some plants even have video screens in break rooms, lunch rooms, visitor lounges, and hallways that display many kinds of safety messages, conveniently controlled by user-friendly computer software.

Who determines the content of these messages? I know who should—the target audience for these signs. The same people expected to follow the specific behavioral advice should have as much input as possible in defining message content. Many organizations can get suggestions for safety messages just by asking. But if employees are not accustomed to giving safety suggestions, they might need a positive consequence to motivate their input.

WORKER-DESIGNED SAFETY SLOGANS

This is what I'm talking about: In 1985, employees and visitors driving into the main parking lot for Ford World Headquarters in Dearborn, Michigan, passed a series of four signs arranged with sequential messages, like the old Burma Shave signs. The messages were rotated periodically from a pool of 55 employee entries in a limerick contest for safety-belt promotion. My three favorites are illustrated in Figure 9.5.

Note that the last sign in each series of safety-belt promotion messages at Ford World Headquarters includes the name and department of the author. This public recognition, with the author's permission, of course, provides a positive consequence or reward to the participant. It also reminds all sign viewers that many different people from various work areas are actively involved in safety. Through positive recognition and observational learning,

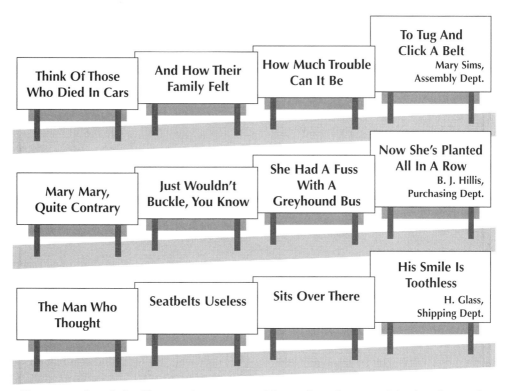

Figure 9.5. Sample buckle-up activators created by Ford employees and displayed at Ford World Headquarters in 1986.

including vicarious reinforcement, this simple technique promotes ownership and involvement in a safety process. This leads to the next principle: Involve the target audience.

Principle #4: Involve the Target Audience

Figure 9.6 depicts a sample promise card for involving people in a commitment to perform a particular behavior. The target behavior to increase in frequency could be selected by a safety director, group leader, or through a group consensus discussion. This behavior is written on the promise card, perhaps by each individual in a group. Group members decide on the duration of the promise period, and write the end date on the card. Then each group member should be encouraged, not coerced, to sign and date their cards. I have found this group application of the safe behavior promise strengthens a sense of group cohesion or belongingness. Follow these procedural points for optimal results.

- Define the behavior specifically
- Involve the group in discussing the personal and group value of the behavior

Figure 9.6. A sample promise card for activating a behavioral commitment.

- Make the commitment for a specified period of time that is challenging but not overwhelming
- Assure everyone that signing the card is only a personal commitment, not a company contract
- There should be no penalties (not even criticism) for breaking a promise
- Encourage everyone to sign the card, but do not use pressure tactics
- Each signer should keep the promise card in his or her possession, or post it in their work area as a reminder

The more involvement and personal choice solicited during the completion of this activator strategy, the better each individual feels about the process. Personal commitment to perform a specific behavior is activated as a result; and those involved in the process should feel obligated to fulfill the promise. Signing the card publicly in a group meeting also implicates social consequences to motivate compliance. That is, many participants will be motivated to keep their promise to avoid disapproval from a group member. Plus, when individuals keep their promise, recognition and approval from the group reinforces and supports maintenance of the targeted safe behavior.

THE AIRLINE LIFESAVER

And now, another personal experience with an activator intervention, one that I've used since November, 1984, whenever boarding a commercial airplane. I hand the flight attendant a 3″ × 5″ Airline Lifesaver card. The card indicates that airlines have been the most effective promoters of seat-belt use, and requests that someone in the flight crew make the following announcement: *Now that you have worn a seat belt for the safest part of your trip; the flight crew would like to remind you to buckle up during your ground transportation.*

From November, 1984, to February, 1995, I distributed the Airline Lifesaver on 620 flights, and on more than 40 percent of these occasions the flight attendant gave a public buckle-up reminder. Many friends have laughed at the Airline Lifesaver, claiming I'm wasting my time. A common comment: "No one listens to the airline announcements anyway. Besides, do you really think that would be enough to motivate people to buckle up if they don't already?"

Ah, but consider this personal experience from the mid-1980s. I observed a woman approach the driver of an airport shuttle, asking her to "Please use your safety belt." The driver immediately buckled up. When I thanked the woman for making the buckle-up request, she replied that she normally wouldn't be so assertive but she had just heard a buckle-up reminder on her flight, "and if a stewardess can request safety-belt use, so can I."

Except for a few anecdotes like this one, it's impossible to assess the direct buckle-up influence of the Airline Lifesaver. But it's "safe" to assume that the beneficial, large-scale impact of this activator is a direct function of the number of individuals who deliver the reminder card to airline personnel. If the delivery of an Airline Lifesaver does not influence a single airline passenger to use a safety belt during ground transportation, at least the act of handing an Airline Lifesaver card to another person should increase the card deliverer's commitment to personal safety-belt use.

Of course, the primary purpose of getting involved in a safety intervention is to prevent injury or improve a person's quality of life. Unfortunately, we rarely see these most important consequences. Thus, we need motivation: feedback, interpersonal approval, and self-talk. We tell ourselves that the safe behavior is "the right thing to do," and that someday an injury will be prevented. We can't count the number of injuries we prevent; we just need to "keep the faith."

On December 28, 1994, I received a special letter from Steven Boydston, assistant vice president of Alexander & Alexander of Texas, Inc., which helps me "keep the faith" that the Airline Lifesaver and other activator interventions for safety do make a difference. The motivating words in this letter are repeated in Figure 9.7. This success story is itself an activator for such proactive interventions as the Airline Lifesaver. It sure worked for me.

THANKS!!! On December 11, 1994 I was a passenger on Flt. 499 from Houston to San Francisco. At the end of the flight the pilot came on the speaker and said: "Now that the safest part of your journey is over and you are about to make the most dangerous part of your journey, please remember to use your seat belt...." I may have the words a little off, but I think you know the message. This was the second time I heard the message. The first was at the ASSE PDC in Dallas.

I am an obsessive seat belt user but when I got into the taxi, the seat belt was, as usual, buried in the seat. Usually when I find this in a cab I say, "the heck with it," but I can honestly say the pilot's message motivated me and I "dug out" the seat belt.

At over 70 mph the taxi hydroplaned and struck the guardrail. Thank God for the new barriers that prevent cars from being thrown back into traffic. Thank you and the pilot for the reminder. My wife and children are also grateful. I suffered a neck and shoulder injury—I think it is relatively minor. It could have been much worse. I should mention the driver was O. K. He was wearing his seat belt.

In safety it is seldom we can point to a particular event we can take credit for. This is one for you. Thanks again.

Figure 9.7. The encouraging letter I received from Steven Boydston on December 28, 1994.

Principle #5: Activate Close to Response Opportunity

Most of the effective activators discussed so far occurred at the time and place the target behavior should happen. I give airline attendants the Airline Lifesaver card when boarding the plane. Actually, I believe I would get more compliance with the request for a buckle-up announcement if I handed an attendant the announcement card at the end of the flight—closer to the opportunity to make the requested response. In fact, when I inquire about the lack of a buckle-up announcement while deplaning, some flight attendants tell me they forgot about the card.

BUCKLE-UP ROAD SIGNS

Over a two-year period my students evaluated the behavioral impact of buckle-up activators located along the road in my hometown of Newport—a small rural community in southwest Virginia. We started collecting baseline data in March, 1993, by unobtrusively observing and recording the safety belt use of vehicle drivers and passengers from a parked vehicle near the intersection of a four-lane highway (Highway 460) and the two-lane road (Route

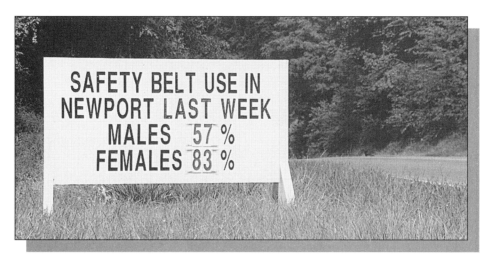

Figure 9.8. The feedback sign located near the intersection of Highway 460 and Route 42 into Newport, Virginia.

42) leading into Newport. Observations were taken of vehicles entering or leaving Route 42 to Newport, as well as vehicles continuing on Highway 460, during most weekdays from approximately 4:00 to 6:30 p.m., when the Newport traffic was heaviest.

After 13 weeks of baseline observation, the sign shown in Figure 9.8 was positioned approximately 7 feet from Route 42 and 300 feet from the intersection of Route 42 and Highway 460. The sign was 8' × 4' and the buckle-up message shown in Figure 9.8 was painted on both sides in black 8″ high letters against a white background. The sign could not be seen by occupants of vehicles continuing along Highway 460.

Vehicle observations continued for 24 weeks, then the feedback sign was removed. After 21 weeks of observation during this withdrawal condition, the signs were reinstated, but with a different message. We wanted to see if safety-belt use could be activated with a sign that did not need to be changed weekly to reflect belt-use feedback. The new message was, "We Buckle Up in Newport to Set An Example for Our Children." Note that this message specifies an actively caring (example setting) consequence for the requested behavior. We continued to record safety belt use for 19 more weeks while this new buckle-up activator was in place.

The results of our long-term field observations show quite clearly that both signs increased safety belt use substantially. While mean safety-belt use in vehicles traveling on Highway 460 remained relatively stable, the mean safety belt use in vehicles entering or exiting the road on which the signs were placed increased by almost 30 percent with placement and removal of the buckle-up activators. This suggests that large-scale increases in safety belt use would occur if communities and companies nationwide implemented this simple activator intervention. Our findings also suggest that an activator message referring to actively caring consequences can be as effective as a feedback sign that requires more effort to implement due to the need to collect behavioral data and post weekly feedback.

Finally, it's instructive to note that our second activator intervention ended abruptly when vandals carried the 70-pound sign about 100 yards and threw it in my pond. Consider this: We did not solicit community approval or involvement when developing or implementing this intervention. We just built the signs and put them in place. It's possible, even likely, that this apparent one-person decision to post a community sign irritated some residents. A few were so outraged that they reacted with countercontrol, perhaps to gain a sense of freedom from our obvious attempt to control their behavior.

We will return to this issue again in Chapter 14 when I discuss ways to increase perceptions of personal control and empowerment to boost involvement in efforts to achieve a Total Safety Culture.

Principle #6: Implicate Consequences

Field research shows the effective influence of salient activators implemented in close proximity to an opportunity to perform the specified target behavior. It's important to realize, though, that the target behaviors were all relatively convenient to perform. We're talking about depositing handbills in a particular receptacle, choosing certain products, using available safety glasses and safety belts. There is plenty of evidence that activators alone won't succeed when target behaviors require more than a little effort or inconvenience.

Many of the activator strategies illustrated in this chapter were explicitly or implicitly connected to consequences. Signing a promise card or public declaration, for example, implicates social approval versus disapproval for honoring versus disavowing a commitment. Consequences motivated employees to create safety slogans, and the most influential activators usually made reference to consequences. In a similar vein, the salient beep of a radar detector effectively motivates reduced vehicle speeds because it enables drivers to avoid a negative consequence—an encounter with a police officer.

INCENTIVES VERSUS DISINCENTIVES

Activators that signal the availability of a consequence are either incentives or disincentives. An *incentive* announces to an individual or group, in written or oral form, the availability of a reward. This pleasant consequence follows the occurrence of a certain behavior or an outcome of one or more behaviors. In contrast, a *disincentive* is an activator announcing or signaling the possibility of receiving a penalty. This unpleasant consequence is contingent on the occurrence of a particular undesirable behavior.

The next chapter discusses how to design and apply consequences to motivate behavior. At this point, it's important to understand that *the power of an activator to motivate behavior depends on the consequence it signals.* Figure 9.9 illustrates this connection between activator and consequence. If a sign like the one shown in Figure 9.9 motivated a driver to attempt safer driving practices, it would work due to the potential consequences implied by the activator. Every time the driver got into the vehicle, she would be reminded of potential consequences for

Figure 9.9. The most powerful activators imply consequences.

certain driving practices. Incidentally, do you perceive the sign on the vehicle in Figure 9.9 as an incentive or a disincentive?

My guess is you perceived it as a *disincentive.* I'll bet you saw the sign as a threat to reduce at-risk driving, rather than an incentive to encourage safe driving. You wouldn't expect dad to get a phone call commending his daughter's driving. If any phone calls were made, they would be to criticize at-risk or discourteous driving. This is exactly how my daughter, Krista, perceived the activator in Figure 9.9. As the illustration suggests, she flatly refused to drive with such a sign on our car.

Much to Krista's chagrin, I actually painted and mounted the sign, but the only benefit from that effort was my daughter's increased motivation to work with the critical behavior checklist described in Chapter 8. "Let's lose this sign, dad," she asserted, "and focus on giving each other feedback with the checklist." I was happy to comply. This exercise simply reminded me that we're socialized to expect more negative consequences for our mistakes than positive consequences for our successes. The next chapter explores this unfortunate reality in greater detail.

SETTING GOALS FOR CONSEQUENCES

Let's talk about safety goals in the context of activators that imply consequences. Included among Dr. W. Edwards Deming's 14 points for quality transformation are "eliminate slogans, exhortations, and targets for the work force...eliminate work standards...management by objectives, and management by the numbers" (Deming, 1985, p 10.). Does this mean we should stop setting safety objectives and goals? Should we stop trying to activate safe behaviors with signs, slogans, and goal statements? Does this mean we should stop counting OSHA recordables and lost-time cases, and stop holding people accountable for their work injuries?

Answers to all of these questions are "yes," if you take Deming's points literally. However, my evaluation of Dr. Deming's scholarship and workshop presentations, and my personal communications with Dr. Deming in 1990 and 1991, have led me to believe that Deming meant we should eliminate goal setting, slogans, and work targets as they are currently implemented. Dr. Deming was not criticizing appropriate use of goal setting, management by objectives, and activators; rather he was lamenting the frequent incorrect use of these activator interventions.

SET SMART GOALS. I remember the techniques for setting effective goals with the acronym SMART, as illustrated in Figure 9.10. SMART goal setting defines what will happen when the goal is reached (the consequences), and tracks progress toward achieving the goal. Rewarding feedback from completing intermediate steps toward the ultimate goal is a consequence that motivates continued progress. Of course, it's critical that the people asked to work toward the goal "buy in" or believe in the goal, and believe they have the skills and resources to achieve it.

FOCUS ON THE PROCESS. Safety goals should focus on process activities that can contribute to injury prevention. Workers need to discuss what they can do to reduce injuries, from reporting and investigating near hits to conducting safety audits of environmental conditions and work practices. One safety steering committee I worked with wanted to increase daily interpersonal communications regarding safety. They set a goal for their group to achieve 500 safety communications within the following month. To do this, they had to

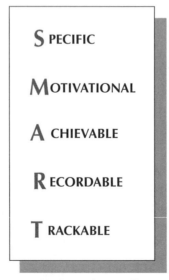

S PECIFIC

M OTIVATIONAL

A CHIEVABLE

R ECORDABLE

T RACKABLE

Figure 9.10. SMART goals are effective activators.

develop a system for tracking and recording "safety talk." They designed a wallet-sized "SMART Card" for recording their interactions with others about safety. One member of the group volunteered to tally and graph the daily card totals.

Another work group I consulted with recently set a goal of 300 behavioral observations of lifting. Employees had agreed to observe each other's lifting behaviors according to a critical behavior checklist they had developed. If each worker completed an average of one lifting observation per day, the group would reach their goal within the month. Each of these work groups reached their safety goals within the expected time period, and as a result they celebrated their "small win" at a group meeting, one with pizza and another with jelly-filled donuts. Perhaps subsequent group goals should focus on healthy diet choices!

These two examples illustrate the use of SMART goals, and depict safety as process-focused and achievement-oriented, rather than the standard and less effective outcome-focused and failure-oriented approach promoted by injury-based goals. More importantly, these goals were employee driven. Workers were motivated to initiate the safety process because it was their idea. They got involved in the process and owned it. And they stayed motivated because the SMART goals were like a roadmap telling them where they were going, when they would get there, and how to follow their progress along the way. This is much more effective than the typical top-down method for setting goals.

In Conclusion

In this chapter I have presented examples of intervention techniques called activators. They occur before desired or undesired behavior to direct potential performers. Based on rigorous behavioral science research and backed by real-world examples, six principles for maximizing effective activators were given:

- Specify behavior
- Maintain salience with novelty
- Vary the message
- Involve the target audience
- Activate close to response opportunity
- Implicate consequences

We're constantly bombarded with activators. At home we get telephone solicitations, junk mail, television commercials, and verbal requests from family members. At work it's phone mail, e-mail, memos, policy pronouncements, and verbal directions from supervisors and coworkers. On the road there's no escape from billboards, traffic signals, vehicle displays, radio ads, and verbal communication from people inside and outside our vehicles. As discussed in Chapter 5, we selectively attend to some of these activators, ignoring others. Only a portion of the activators we perceive actually influences our behavior. Understanding the six principles discussed in this chapter can help you predict which ones will influence behavior change.

Obviously, we don't need more activators in our lives. We certainly do need more effective activators to promote safety and health. It would be far better to make a few safety activators more powerful than to add more activators to a system already overloaded with information. We need to plan our safety activators carefully so the right safety directives receive the attention and ultimate action they deserve.

If you want an activator to motivate action, you need to imply consequences. The most powerful activators make the observer aware of consequences available following the performance of a target behavior. Consequences can be positive or negative, intrinsic or extrinsic to the task, and internal or external to the person. The next chapter will explain the preceding sentence, which is key to getting the most beneficial behavior from an intervention process.

Chapter 10

Intervening with Consequences

Consequences motivate behavior and related attitudes. There are various ways this happens. Consequences can be positive or negative, intrinsic (natural) or extrinsic (extra) to a task, and internal or external to a person. These characteristics need to be considered when designing and evaluating intervention programs. This chapter explains why, and provides principles and practical procedures for motivating people to work safely over the long term. I'll show you how to influence behavior and attitudes so that both are consistent with a Total Safety Culture.

People do what they do because of what happens to them when they do it.

This quote from Aubrey Daniels (1994, p. 25) represents a key principle of human motivation and behavior-based safety. Although supported by substantial research (Skinner, 1938), it actually runs counter to common sense.

Think about it: When people ask us why we did something, we're apt to say, "I wanted to do it," or "I was told to do it," or "I needed to do it." These explanations sound as if the cause of our behavior comes before we act. This perspective is supported by numerous "pop psychology" self-help books and audiotapes that say people motivate themselves with positive self-affirmations or optimistic thinking and enthusiastic intentions. In other words, behavior is caused by some external request, order, or signal; or by an internal force, drive, desire, or need.

This chapter will explain the fallacy in this common-sense argument and show ways to maximize the impact of an extrinsic reward process. Again, the research-supported principle is that activators direct behavior, and consequences motivate behavior. But the type of consequence certainly influences the amount of motivation, as this chapter will explain.

The Power of Consequences

Popular author and humorist Robert Fulghum (1988) wrote *All I Really Need to Know I Learned in Kindergarten,* claiming he learned all the basic rules or norms for socially acceptable adult behavior as a young child. The list of rules in Figure 10.1 was excerpted from Mr. Fulghum's famous book. Rules like *share everything, play fair, don't fight,* and *clean up your own mess* were taught to most of us early on. These are clearly ideal edicts to live by. Perhaps you still recall a teacher or parent using these rules to try to shape your behavior. Did it work? Do you follow each of these basic norms regularly, for no other reason or consequence except your realization that it's the right thing to do? Imagine what a better world we'd live in if everyone followed the simple rules listed in Figure 10.1 from a self-directed, principle-focused perspective. Alas, there are signs everywhere that this isn't so.

The last two kindergarten rules in Figure 10.1 are directly relevant to safety, and in fact reflect basic themes of this text. As discussed previously, especially in Chapter 9, safety requires people to stick together in a spirit of shared belonging and interdependence. But sometimes we need activators to remind us of this critical rule, and consequences to keep us working together for safety.

"LOOK," Fulghum's last rule, is key to behavior-based safety and to achieving a Total Safety Culture. This implies the "defensive working style" employees need to adopt. In a Total Safety Culture, everyone looks for ways to improve safety by intervening to reduce at-risk behaviors and increase safe behaviors. In Chapter 9 we discussed ways to intervene with activators. Here we focus on the more powerful intervention approach—manipulating consequences.

❏ Share everything
❏ Play fair
❏ Don't hit others
❏ Put things back where you found them
❏ Clean up your own mess
❏ Don't take things that aren't yours
❏ Say you're sorry when you hurt someone
❏ Wash your hands before you eat
❏ Flush
❏ When you go out into the world, watch out for traffic, hold hands, and stick together
❏ And remember the Dick-and-Jane books and the first word you learned—the biggest word of all: **LOOK**

Figure 10.1. Basic rules of social life we learned well as children, but do not necessarily follow as adults. (Excerpted from Fulghum, 1988)

INTRINSIC VERSUS EXTRINSIC CONSEQUENCES

Most applied behavioral scientists view *intrinsic motivation* rather differently than the description used in pop psychology books. The behavior-based perspective is supported by research and our everyday experience. Plus, it's objective, practical, and useful for developing situations and programs to motivate behavior change.

Simply put, "intrinsic" does not mean "inside" people, where it cannot be observed, measured, and directly influenced. Rather, "intrinsic" refers to the nature of the task in which an individual is engaged. Intrinsically motivated tasks, or behaviors, lead naturally to external consequences that support the behavior (rewarding feedback) or give information useful for improving the behavior (corrective feedback).

Most athletic performance, for example, includes natural or "intrinsic" consequences that give rewarding or correcting feedback. These consequences, intrinsic to the task, tell us immediately how well we have performed at swinging a golf club, shooting a basketball, or casting a fishing lure, for example. They motivate us to keep trying, sometimes after adjusting our behavior as a result of the natural feedback directly related, or intrinsic, to the task.

Take a look at the fisherman in Figure 10.2. Some psychologists would claim he is motivated from within, or self-motivated. They use the term "intrinsic motivation" to refer to this state. In contrast, the behavioral scientist points to the external consequences naturally motivating the fisherman's behavior. These cause him to focus so completely on the task at hand that he's not aware of his wife's mounting anger—or he's ignoring her. He may also be unaware that his supply of fish is creating a potential hazard. In a similar way, safety can be compromised because of excessive motivation for production. Rewards intrinsic to production can cause this motivation.

Note that the "worker" in this picture does not receive a reward for every cast. In fact, he's on an intermittent reinforcement schedule. He catches a fish once in a while. This kind of reward schedule is most powerful in maintaining continuous behavior. Anyone who has gambled understands. Some say gambling is a disease, when in fact gambling is behavior maintained by intermittent rewarding consequences.

JUST ONE MORE FISH - THAT'S WHAT YOU SAID YESTERDAY.

Figure 10.2. Some tasks are naturally motivating because of intrinsic consequences.

Some tasks do not provide intrinsic or natural feedback. In this case, it's necessary to add an extrinsic, or extra, consequence to support or redirect the behavior. Many, if not most, safety behaviors fall in this category.

In fact, many safety practices have intrinsic *negative* consequences, such as discomfort, inconvenience, and reduced pace, that naturally discourage their occurrence. Thus, there is often a need for extrinsic supportive consequences, like intermittent praise, recognition, novelties, and credits redeemable for prizes, to shape and maintain safe behaviors. The intent is not to control people, but to help people control their own behavior by offering positive reasons for making the "safe" choice.

INTERNAL VERSUS EXTERNAL CONSEQUENCES

The intrinsic and extrinsic consequences discussed so far are external to the individual. In other words, they can be observed by another person. Behavioral scientists focus on these types of consequences to develop and evaluate motivational processes because they can be objective and scientific when dealing with external, observable aspects of people.

But behavioral scientists do not deny the existence of internal factors that motivate action. There is no doubt that we talk to ourselves before and after our behaviors, and this self-talk influences our performance. We often give ourselves internal verbal instructions, called intentions, before performing certain behaviors. After our activities we often evaluate our performance with internal consequences. In the process we might motivate ourselves to press on (with self-commendation) or to stop (with self-condemnation).

When it comes to safety and health, internal consequences to support the right behavior are terribly important. Remember, external and intrinsic (natural) consequences for safe behaviors are not readily available, and we can't expect to receive sufficient support (extra consequences) from others to sustain our proactive, safe, and healthy choices. So we need to talk to ourselves with sincere conviction to boost our intentions, and with genuine self-reinforcement after we do the right thing. When we receive special external consequences for our efforts, we need to savor these and use them later to bolster our self-reinforcement.

Managing Consequences for Safety

At this point, I'm sure you appreciate the special message reflected in Figure 10.3. Submitting safety suggestions is an activity not typically followed by motivating consequences. In many work cultures the idea of safety suggestions has long since passed. The suggestion boxes are empty. Does this mean there are no more good suggestions? Is the work force not creative enough? You know the

Figure 10.3. Some tasks require supportive consequences.

answer to both of these questions is a resounding "No."

Let me give you an example. I recently worked with safety leaders at a Toyota Motor manufacturing plant in Georgetown, Kentucky, whose 6,000 employees submitted more than 35,000 quality, production, or safety-related suggestions in 1994. And a greater number of suggestions were expected in 1995. Many employees in this culture are motivated internally to submit suggestions, but external consequences are in place to keep the process going. Employees receive timely feedback regarding the utility and feasibility of every suggestion, and if the suggestion is approved, they are empowered to implement it themselves. Also, the individual or team responsible receives 10 percent of the savings for the first year that result from the implemented suggestion. Such external, extra and meaningful—in this case economic—consequences motivate a large work force to make a difference.

THE CASE AGAINST NEGATIVE CONSEQUENCES

To subdue influences supporting at-risk behavior, it's often tempting to use a punisher or penalty. All that's needed is a policy statement or some type of top-down mandate specifying a soon, certain, and sizable negative consequence following specific observable risky behaviors. Couldn't this contingency be powerful enough to override the many natural positive consequences for taking risks?

Yes, behavioral scientists have found negative consequences can permanently suppress behavior if they are severe, certain, and immediate. But before using "the stick," you should understand the practical limitations and undesirable side effects of using negative consequences to influence behavior.

ESCAPE. Animals and people attempt to avoid situations with a predominance of negative consequences. Sometimes this means staying away from those who administer the punishment. Humans will often attempt to escape from negative consequences by simply "tuning out," or perhaps cheating or lying. Of course the ultimate escape from excessive negative consequences is suicide. Indeed, it's not uncommon for an individual to commit suicide in order to escape control by aversive stimulation, which can include the intractable pain of an incurable disease, physical or psychological abuse from a family member, or perceived harassment by an employee or coworker.

Figure 10.4. Fear of negative consequences is motivating.

Unpleasant attitudes or emotional feelings are produced when people work to escape or avoid negative consequences. As shown in Figure 10.4, negative consequences can influence behavior dramatically, but such situations are usually unpleasant for the "victim." Under fear arousal conditions, people will be motivated to do the right thing, but only when they have to. They feel controlled, and as discussed in Chapter 6, this can lead to distress and burnout. Obviously, this type of contingency and side effect is incompatible with a Total Safety Culture where people feel in control, and are ready and willing to go beyond the call of duty for another person's safety and health.

AGGRESSION. Instead of escaping, people might choose to attack those perceived to be in charge. For example, murder in the workplace is rapidly increasing in the United States; and the most frequent cause appears to be reaction or frustration to control by negative means. But aggressive reaction to this kind of control might not be directed at the source.

An employee frustrated by top-down aversive control at work might not assault his boss directly, but rather slow down production, sabotage a safety program, steal supplies, or vandalize industrial property. Or the employee might react with spousal abuse. Then the abused spouse might react by slapping a child. The child, in turn, might punch a younger sibling. And the younger sibling might punch a hole in a wall or kick the family pet—all as a result of perceived control by negative consequences.

APATHY. Apathy is a generalized suppression of behavior. In other words, the negative consequences not only suppress the target behavior but might also inhibit the occurrence of desirable behaviors. Regarding safety, this could mean a decrease in employee involvement. When people feel controlled by negative consequences, they are apt to simply resign themselves to doing only what's required. Going beyond the call of duty for a coworker's health or safety is out of the question.

COUNTERCONTROL. No one likes feeling controlled, and situations that influence these feelings in people do not encourage buy-in, commitment, and involvement. In fact, some people only follow top-down rules when they believe they can get caught, as typified by drivers slowing down when noticing a police car. Some people look for ways to beat the system they feel is controlling them, so you have vehicle drivers purchasing radar detectors. This is an example of "countercontrol," the fourth undesirable side effect of negative consequence contingencies.

I met an employee once who exerted countercontrol by wearing safety glasses without lenses; when wearing his "safety frames," he got attention and approval from certain coworkers. Perhaps these coworkers were rewarded vicariously when seeing him beat the system they perceived was controlling them also.

Figure 10.5. Countercontrol behavior is usually directed at those in charge of negative consequences.

Figure 10.5 illustrates an example of countercontrol. Although the supervisors might view the behavior as "feedback," it is countercontrol if it occurred to regain control or assert personal freedom. A perceived loss of control or freedom is most likely when a negative consequence contingency is implemented. Also, countercontrol behavior is typically directed at those in charge of the negative consequences.

DISCIPLINE AND INVOLVEMENT

Let's specifically discuss traditional discipline for safety—a form of top-down control with negative consequences. I've met many managers who include a "discipline session" as part of the corrective action for an injury report. The injured employee gets a negative lecture from a manager or supervisor whose safety record and personal performance appraisal were tarnished by the injury.

These "discipline sessions" are unpleasant for both parties, and certainly do not encourage personal commitment or buy-in to the safety mission of the company. Instead, the criticized and embarrassed employees are simply reminded of the top-down control aspects of corporate safety, usually resulting in increased commitment not to volunteer for safety programs, nor to encourage others to participate. In this case, the culture loses the involvement of invaluable safety participants.

PROGRESSIVE DISCIPLINE. Whenever I teach behavior management principles and procedures, the question of how to deal with the repeat offender frequently comes up. Aren't there times when discipline is necessary? Doesn't an individual who breaks the rules after repeated warnings or confrontations deserve a penalty? Through progressive discipline these individuals receive top-down penalties, starting with a verbal warning, then written warnings, and eventually dismissal. In some cases, dismissal is indeed the best solution for uncooperative individuals who can be a divisive and dangerous factor in the workforce.

But keep in mind that progressive discipline can seriously damage your efforts to achieve a Total Safety Culture. Aggression or countercontrol moves from these individuals can not only decrease employee trust and involvement, but put employees at risk. The disgruntled worker might neglect a safety precaution. Also, escalating penalties for misbehavior can progressively turn off an employee who might have become a true supporter, even a leader, of the safety effort if initially approached differently. Thus, involvement and traditional discipline don't mix, even when the discipline is progressive.

ALTERNATIVES TO PROGRESSIVE DISCIPLINE. So what are alternatives for dealing with repeat safety offenders? One approach is to focus on the positive and give rewarding feedback for safe behavior. Then when corrective feedback is warranted it will be accepted if given appropriately. But what about the employee who seemingly does little to deserve rewarding feedback? Well, anyone can improve from their own baseline level, and improvement can be the target for rewarding feedback.

What about employees who commit serious safety violations which put themselves or, worse yet, *others* at risk, such as not complying with standard lock out/tag out procedures? Isn't it necessary to make a visible and immediate statement about such acts, and set an example that such infractions cannot be tolerated? Yes, indeed. But what kind of disciplinary action will have beneficial and lasting impact, without decreasing involvement and increasing countercontrol? I suggest that developing a "Total Safety Culture Empowerment Council" is the next step.

EMPLOYEE EMPOWERMENT COUNCIL. If a student at my university is caught cheating by an instructor or another student, his or her name is submitted to the "University Honor Council" along with details about the incident. Students volunteer to serve on the honor council, and a "Chief Justice" is elected by the entire student body. University faculty or staff only get involved in this discipline system when making a referral or when presenting evidence during the honor council's fact-finding mission.

After fact finding and deliberating, the honor council might dismiss the case, recommend a penalty for the alleged cheater, and/or suggest changes in the instructor's procedures or policies. In one case, an instructor was given advice on the use of different test forms and classroom seating arrangements. In another case, a professor was advised to eliminate his "closed book, take home" exams. The rationale behind the university honor council is that those most affected by cheating and those most capable of gathering and understanding the facts about alleged cheating should run the system. This disciplinary system is administered for and by the students it serves.

Given that employees typically have the most direct influence over their peers, and that top-down discipline usually decreases voluntary involvement in desirable safety processes, the idea of a TSC Empowerment Council seems logical and intuitively appealing. If a council of people representative of the entire work force serves the fact-finding and corrective-action functions of safety discipline, employee involvement would be enhanced rather than hurt by a discipline system. Such a council could offer the guidance, leadership, and counseling implied by the Latin roots of discipline: *disciplina*, meaning instruction or training, and *discipulus*, referring to a learner. *Disciple* is also derived from the same Latin roots.

Dos and Don'ts of Safety Rewards

Now let's look at the flip side of discipline: rewards. In one-on-one situations with children at home or in school, using positive consequences to increase desirable behavior is straightforward and easy. But using rewarding consequences effectively with adults in work settings is easier said than done, especially when it comes to safety. Throughout my 30+ years of professional experience in motivational psychology, I've seen more inappropriate reward programs in occupational safety than in any other area. This is unfortunate, because the effective use of extra positive consequences is often critically important to overcome the readily available influences supporting risky behavior.

By this point, I'm sure you understand the difference between an incentive and a reward. An incentive is an activator that promises a particular positive consequence (a reward) when a correct behavior occurs. Disincentives, on the other hand, are activators such as rules and policies that announce penalties for certain undesired behavior. Remember, the motivating power of incentives and disincentives depends on following through. Rules or policies that are not consistently and justly enforced with penalties for noncompliance are often disregarded. If promises of rewards are not fulfilled when the behaviorial criteria are reached, subsequent incentives might be ignored. This need for consistent delivery of consequences—whether positive or negative—makes it quite challenging to develop and manage an effective motivational program for safety.

DOING IT WRONG

Most incentive/reward programs for occupational safety do not specify behavior: Employees are rewarded for avoiding a work injury or achieving a certain number of "safe work days." So what behavior is motivated? Not reporting injuries.

If having an injury endangers your reward or, worse, the reward for an entire work group, there is pressure to avoid reporting that injury, if possible. Many of these non-behavioral, outcome-based incentive programs involve substantial peer pressure because they use a group-based contingency. That is, if anyone in the company or work group is injured, everyone loses their reward. Not surprisingly, I've seen coworkers cover for an injured employee in order to keep accumulating "safe days."

These incentive programs might decrease the numbers of reported injuries, at least over the short term, but corporate safety is obviously not improved. Indeed, such programs often create apathy or helplessness regarding safety achievement. Employees develop the perspective that they can't really control their injury record, but must cheat or beat the system to celebrate the "achievement" of an injury reduction goal.

DOING IT RIGHT

Here are seven basic guidelines for establishing an effective incentive/reward program that manages the human element of industrial health and safety:

1. The behaviors required to achieve a safety reward should be specified and perceived as achievable by the participants.

2. Everyone who meets the behavioral criteria should be rewarded.

3. It's better for many participants to receive small rewards than for one person to receive a big reward.

4. The rewards should be displayed and represent safety achievement. Coffee mugs, hats, shirts, sweaters, blankets, or jackets with a safety message are preferable to rewards that will be hidden, used, or spent.

5. Contests should not reward one group at the expense of another.

6. Groups should not be penalized or lose their rewards for failure by an individual.

7. Progress toward achieving a safety reward should be systematically monitored and publicly posted for all participants.

Guideline 2 recommends against the popular lottery or raffle drawing. As illustrated in Figure 10.6, a lottery results in one "lucky" winner being selected and a large number of "unlucky" losers. The announcement of a raffle drawing might get many people excited, and if lottery tickets are dispensed for specific safe behaviors, there is some motivational benefit. Eventually, however, the valuable reward is received by a lucky few.

Also, I perceive a disadvantage in linking chance with safety. It's bad enough we use the word "accident" in the context of safety processes. The

Figure 10.6. Raffle drawings result in few "lucky" winners and many "unlucky" losers, and thus can have undesirable side effects.

first definition of "accident" in my *New Merriam-Webster Dictionary* (1989) is "an event occurring by chance or unintentionally" (p. 23). We shouldn't add to this inference that injuries are chance occurrences.

I've worked with a number of safety directors who used a lottery incentive program and vowed they would never do it again. The big raffle prize, such as a snowmobile, pick-up truck or television set, is displayed in a prominent location. Everyone gets excited—temporarily—about the possibility of winning. But their attention is directed at the big prize instead of the real purpose of the program: to keep everyone safe. *The material rewards in an incentive program should not be perceived as the major payoff.* Incentives are only reminders to do the right thing, and rewards serve as feedback and a token of appreciation for doing the right thing.

More important than external rewards is the way they are delivered. Rewards should not be perceived as a means of controlling behavior, but as a declaration of sincere gratitude for making a contribution. If many people receive this recognition, you have made deposits in the emotional bank accounts of potential actively caring participants in a Total Safety Culture. That's why it's better to reward many than few (Guideline 3).

When rewards include a safety logo or message (Guideline 4), they become activators for safety when displayed. Also, if the safety message or logo was designed by representatives from the target population, the reward takes on special meaning (as discussed previously in Chapter 9). Special items like these cannot be purchased anywhere, and from the perspective of internal consequences, they are more valuable than money.

As depicted in Figure 10.7, contests that pit one group against another can lead to an undesirable win-lose situation (Guideline 5). Safety needs to be perceived as win-win. This means developing a contract of sorts between each employee that makes everyone a stockholder in achieving a Total Safety Culture. Everyone in the organization is on the same team. Team performance within departments or work groups can be motivated by providing team rewards or bonuses for team achievement. Every team that meets the "bonus" criteria should be eligible for the reward. In other words, Guideline 2 should be applied when developing incentive/reward programs to motivate team performance.

Figure 10.7. Safety contests can motivate unhealthy competition.

Obviously, developing and administering an effective incentive/reward program for safety requires a lot of dedicated effort. There is no quick fix. But it is worth doing, if you take the time to do it right. As Aubrey Daniels wisely stated, "If you think this is easy, you are doing it wrong" (1994, p. 171). Let's examine an exemplary case study.

AN EXEMPLARY INCENTIVE/REWARD PROGRAM

In 1992, I consulted with the safety steering committee of a Hoechst Celanese company of about 2000 employees to develop a plant-wide incentive program that followed each of the guidelines given above. The steering committee, including four hourly and four salary employees, met several times to identify specific behavior-consequence contingencies. That is, they needed to decide what behaviors should earn what rewards. Their plan was essentially a "credit economy" where certain safe behaviors, which could be achieved by all employees, earned certain numbers of "credits."

At the end of the year, participants exchanged their "credits" for a choice of different prizes, all containing a special safety logo. The variety of behaviors earning credits included: attending monthly safety meetings; special participation in safety meetings; leading a safety meeting; writing, reviewing, and revising a job safety analysis; and conducting periodic audits of environmental and equipment conditions, and certain work practices. For a work group to receive credits for audit activities, the results of environmental and PPE observations had to be posted in the relevant work areas. Only one behavior was penalized by a loss of credits—the late reporting of an injury.

At the start of the new year, each participant received a "safety credit card" for tallying credit earnings each month. Some individual behaviors earned credits for the person's entire work group, thus promoting teamwork and group cohesion. It's noteworthy that this kind of incentive/reward program exemplifies a basic behavior-based principle for health and safety management—observation and feedback. Employees were systematically observed, and they received soon, certain, and positive feedback (a reward) after performing a target behavior. An incentive/reward program is only one of several methods to increase safe work practices with observation and feedback. In the next chapter I address feedback more specifically as an external and extra consequence to prevent injuries.

SAFETY THANK-YOU CARDS

I'd be remiss if I did not describe "Safety Thank-You Cards" in a discussion of exemplary incentive/reward approaches. Over the years, I've seen a wide variety of thank-you cards designed by work teams and used successfully at a number of industrial sites, including plants for Hercules, Hoechst Celanese, Exxon Chemical, Ford, Phillip Morris, General Motors, Logan Aluminum, Westinghouse Hanford Company, Abbott Laboratories, Weyerhaeuser, and M & M Mars.

At some locations thank-you cards were used in a raffle drawing, exchangeable for food, drinks, or trinkets, or displayed on a plant bulletin

Figure 10.8. An actively caring thank-you card used at the Hoechst Celanese plant in Rock Hill, South Carolina.

board as a "safety honor roll." Sometimes the cards could be accumulated and exchanged for tee shirts, caps, or jackets with messages or logos signifying safety achievement. At several plants, the person who delivered a thank-you card returned a receipt naming the recognized employee and describing the behavior earning the consequence, thus creating objective information to define a "Safe Employee of the Month."

At a few locations, the thank-you cards took on a special actively-caring meaning. Specifically, when deposited in a special collection container, each thank-you card was worth 25 cents toward corporate contributions to a local charity or to needy families in the community. The actively caring card used at the Hoechst Celanese plant in Rock Hill, South Carolina, is shown in Figure 10.8. The back of the card included a colorful peel-off symbol which the recognized employee could affix in any number of places as a personal reminder of the recognition. I was surprised but pleased to see a large number of these thank-you stickers on employees' hard hats. Obviously, this actively-caring thank-you approach to safety recognition has great potential as an inexpensive but powerful tool for motivating safe behavior.

In Conclusion

Writing this book was challenging, tedious, overwhelming, tiresome, sacrificing, and exhausting. Observers were apt to say I was self-directed and "intrinsically" motivated. Of course, I know better, and you do too.

Incidentally, I literally wrote the various drafts of this text. I have never learned to type, and therefore have never benefited from the technological magic of computer word processing. My colleagues explain that it's not necessary to be a skilled typist to reap the many "intrinsic" benefits of preparing a manuscript on a computer. "I type slowly with only one finger," some have said, "and still enjoy the wonderful benefits of high-tech computer word processing. I could never go back to preparing a manuscript by hand. You don't know what you're missing."

I'm sure you've noticed my disparate uses of "intrinsic" in the prior paragraphs, and you now understand the two meanings of this popular motivational term. Are my friends and colleagues so enthusiastic about computer-based word processing because of intrinsic (internal) motivation, or because of intrinsic (natural) consequences linked to their computer use? As a review of this chapter, I'm sure you see my point. Word processing on a computer allows for rapid "quick-fix" control of letters, words, sentences, and paragraphs. Computer users also can walk to a printer and obtain a typed "hard copy" of their document for study, revision, or dissemination. All these soon, certain, positive consequences are connected naturally to word-processing behavior. No wonder my friends and colleagues are motivated about computer word processing, and urge me to get on the high-tech "band wagon."

When we earn genuine appreciation and approval from others for what we do, we not only become self-motivated, we also maximize the chances that our activities will influence the behavior of others. In fact, this was my ultimate motivation for soliciting feedback on earlier drafts of this text and for painstakingly refining the presentations. Practice can only improve with feedback. We can only learn to communicate more effectively if we learn how we're coming across to others. If I communicate effectively and earn the approval and appreciation of readers for the principles and procedures presented in this text, injuries and fatalities could be reduced on a large scale. This would be an external and natural consequence of authoring this text—the remote but preeminent motivator for my writing behavior.

Chapter 11
Intervening as a Behavior-Change Agent

This chapter presents the principles and procedures of safety coaching—a key behavior-change process for safety improvement. The letters of **COACH** *represent the critical sequential steps of safety coaching:* **C**are, **O**bserve, **A**nalyze, **C**ommunicate, **H**elp. *This coaching process is clearly relevant for improving behaviors in areas other than safety and in settings other than the workplace.*

The prior two chapters discussed guidelines for developing behavior change interventions. Chapter 9 focused on the use of activators to direct behavior change. Chapter 10 detailed the motivating role of consequences. Several examples employed both activators and consequences. This is applying the three-term contingency (activator-behavior-consequence), and is usually the most influential approach.

Large-scale behavior change is impossible without intervention agents—people willing and able to step in on behalf of another person's safety. In a Total Safety Culture, everyone feels responsible for safety, pursuing it daily. They go beyond the call of duty to identify at-risk conditions and behaviors, and intervene to correct them. This chapter is about becoming a behavior-focused change agent.

In simplest terms, this means observing and supporting safe behaviors or observing and correcting at-risk behaviors. It might involve designing and implementing a particular intervention process for a work team, organizational culture, or an entire community. Or it might mean merely engaging in behavior-focused communication between an observer (the intervention agent) and the person observed. This is safety coaching, and to be effective, certain principles and guidelines need to be followed.

Intervening as a Safety Coach

Coaching is essentially a process of one-on-one observation and feedback. The coach systematically observes the behaviors of another person, and provides behavioral feedback on the basis of the observations. Safety coaches recognize and support the safe behaviors they observe, and offer constructive feedback to reduce the occurrence of any at-risk behaviors. This chapter specifies the steps of safety coaching, points out trainable skills needed to accomplish the process, and illustrates tools and support mechanisms to increase effectiveness.

Figure 11.1. Systematic observation and feedback are key to effective coaching.

The term "coach" is very familiar to us in an athletic context. In fact, winning coaches practice the basic observation and feedback processes needed for effective safety coaching. They follow most of the guidelines reviewed here. As illustrated in Figure 11.1, the most effective team coaches observe the ongoing behaviors of individual players, and record their observations in systematic fashion, using a team roster, behavioral checklist, or videotape. Football coaches, for instance, will spend hours and hours analyzing film. Then they deliver specific and constructive feedback to team members to support or increase desirable behavior and/or to decrease undesirable behavior. Sometimes the feedback is given in a group session, perhaps by critiquing videotapes of team competition. At other times, the feedback is given individually in a personal one-on-one communication. Usually, the one-to-one format has greater impact on individual performance.

The five letters of the word **COACH** can be used to remember the basic ingredients of the most effective coaching—whether coaching the members of a winning athletic team or the individuals in a work group striving for safe behaviors. This is my favorite instructional acronym, because it not only contains the components of an effective coaching process, it lists them in the sequence in which they should occur.

"C" FOR CARE

Caring is the basic underlying motivation for coaching. Safety coaches truly care about the health and safety of their coworkers, and they act on such caring. In other words, they "actively care." When people realize from a safety coach's words and body language that he or she cares, they are more apt to listen to and accept the coach's advice. When people know you care, they care what you know.

OUR EMOTIONAL BANK ACCOUNTS. Stephen Covey (1989) explained the value of interdependence among people—exemplified by appropriate safety coaching—with the metaphor of an "emotional bank account." People develop an emotional bank account with others through personal interaction. Deposits are made when the holder of the account views a particular interaction to be positive, as when they feel recognized, appreciated, or listened to. Withdrawals from a person's emotional bank account occur whenever that individual feels criticized, humiliated, or less appreciated, usually as a result of personal interaction.

Sometimes it is necessary to offer constructive criticism, or even state extreme displeasure with another person's behavior. But if such negative discourse occurs on an "overdrawn or bankrupt account," this corrective feedback will have limited impact. In fact, continued withdrawals from an overdrawn account can lead to defensive or countercontrol reactions. The person will simply ignore the communication or actually do things to discredit the source or undermine the process or system implicated in the communication. Thus, safety coaches need to demonstrate a caring attitude through their personal interactions with others. This maintains healthy emotional bank accounts—operating in the black. The woman in Figure 11.2 is requesting a deposit along with the withdrawal. Our emotional reaction to police officers depends on the proportion of deposits versus withdrawals we have experienced with them.

BUT OFFICER, DON'T I GET CREDIT FOR BUCKLING UP?

Figure 11.2. Our attitude toward police officers would be more positive if we received deposits along with withdrawals.

A SHARED RESPONSIBILITY. People are often unwilling to coach or to be coached for safety because they view safety from an individualistic perspective. To them, it's a matter of individual or personal responsibility. This is illustrated by the verbal expression or internal script, "If Molly and Mike want to put themselves at risk, that's their problem, not mine." For some people a change in personal attitude or perspective is needed in order to motivate coaching. People need to consider safety coaching a shared responsibility to prevent injuries throughout the entire work culture. This requires a shift from an individual to a collective perspective.

Many people accept a collective or team attitude when it comes to work productivity and quality. Coaching for production or quality is part of the job. But coaching for personal safety is often perceived as meddling. People need to understand that safety-related behaviors require as much, if not more, interpersonal observation and feedback as any other job activity.

One way to convince people to accept and support safety coaching as a shared responsibility is to point out their plant's injury record for a certain period of time. While an injury did not happen to them, it did happen to someone, and everyone certainly cares about that. Given this underlying caring attitude, the challenge is to convince others that effective safety coaching by them will reduce injuries to their coworkers. This is enabled by a behavior-based measurement system, as discussed next.

"O" FOR OBSERVE

Safety coaches observe the behavior of others objectively and systematically, with an eye for supporting safe behavior and correcting at-risk behavior. Behavior that illustrates going beyond "the call of duty" for the safety of another person should be especially supported. This is the sort of behavior that contributes significantly to safety improvement and can be increased through rewarding feedback. As illustrated in Figure 11.3, a safety observer does not hide or "spy," and always asks permission first. Only with permission should an observation process proceed.

Observing behavior for supportive and constructive feedback is easy if the coach: a) Knows exactly what behaviors are desired and undesired (an obvious requirement for athletic coaching); and b) takes the time to observe occurrences of these behaviors in the work setting. It is often advantageous—and usually essential—to develop a checklist of safe and at-risk behaviors and to rank them in terms of risk. Ownership and commitment is increased when workers develop their own behavioral checklists.

DEVELOPING A CRITICAL BEHAVIORAL CHECKLIST. Observation checklists can be generic or job-specific. A generic checklist is used to observe behaviors that may occur during several jobs. A job-specific checklist is designed for one job. Deciding which

Figure 11.3. Safety coaches are up-front about their intentions, always asking permission before observing.

items to include on a critical behavior checklist (CBC) is a very important part of the coaching process. A CBC enables coaches to look for critical behaviors. A critical behavior is a behavior that:

- Has led to a large number of injuries or near hits in the past
- Could potentially contribute to a large number of injuries or near hits because many people perform the behavior
- Has previously led to a serious injury or a fatality
- Could lead to a serious injury or fatality

If only a few behaviors will be observed in the beginning, which is often a good way to start a large-scale coaching process, a CBC should be designed for only the most critical behaviors. Several sources can be consulted to obtain behaviors for a CBC, including: a) injury records, b) near-hit reports, c) job hazard analyses, d) standard operating procedures, e) rules and procedural manuals, and f) the workers themselves. People already know a lot about their own safety performance. They know which safety rules they sometimes ignore, and they know when a near hit has occurred to themselves or to others because of at-risk behavior. In addition, it's often useful to obtain advice from the plant doctor, nurse, safety director, or anyone else who maintains injury statistics for the plant.

When starting out, don't develop an exhaustive checklist of critical behaviors. A list can get quite long in a hurry. A long list for one-on-one observations can appear overwhelming, and could inhibit the process. As with anything that is new and needs voluntary support, it's useful to start small and build. With practice, people find a CBC easy to use, and they accept additions to the list. They will also contribute in valuable ways to refine the CBC, from clarifying behavioral definitions to recommending behavioral additions and substitutions. The development and use of a CBC is really a continuous improvement process. Further development and refinement benefits coaching observations, and vice versa.

A work group on a mission to develop a CBC needs to meet periodically to select critical behaviors to observe. I have found the worksheet depicted in Figure 11.4 useful in beginning the development of a CBC. Through interactive discussions, work groups define safe and at-risk behaviors in their own work areas relevant to each category. The category on body positioning and protecting, for example, includes specific ways workers should protect themselves from environment or equipment hazards. This can range from using certain personal protective equipment to positioning their body parts in certain ways to avoid possible injury.

Some categories in Figure 11.4 may be irrelevant for certain work groups, like locking or tagging out equipment or complying with certain permit policies. A work group might add another general procedural category to cover

OPERATING PROCEDURES	SAFE OBSERVATION	AT-RISK OBSERVATION
Body Positioning/Protecting Positioning/protecting body parts (e.g., avoiding line of fire, using PPE, equipment guards, barricades, etc.)		
Visual Focusing Eyes and attention devoted to ongoing task(s)		
Communicating Verbal or nonverbal interaction that affects safety		
Pacing of Work Rate of ongoing work (e.g. spacing breaks appropriately, rushing)		
Moving Objects Body mechanics while lifting, pushing/pulling		
Complying with Lockout/Tagout Following procedures for lockout/tagout		
Complying with Permits Obtaining, then complying with permit(s) (e.g., confined space entry, hot work, exca- vation, open line, hot tap, etc.)		

Figure 11.4. A worksheet for developing a generic critical behavior checklist (CBC).

particular work behaviors. Note that defining safe and at-risk behaviors results in safety training in the best sense of the word. Participants learn exactly what safe behaviors are needed for a particular work process.

A list of work behaviors covering all the generic categories in Figure 11.4 can be extensive and overwhelming. This gives numerous opportunities for coaching feedback, but remember, it takes time and practice to observe behaviors reliably—and to get used to being observed while working. I have found it useful to start the observation procedure with a brief CBC of four or five behaviors, and then build on the list with practice, group discussion, and more practice.

SCHEDULING OBSERVATION SESSIONS. There is no best way to arrange for coaching observations. The process needs to fit the setting and work process. This can only happen if workers themselves decide on the frequency and duration of the observations, and derive a method for scheduling the coaching sessions. I've seen the protocol for effective coaching observations vary widely across plants, and across departments within the same plant. And the success of those processes have not varied as a function of protocol.

The 350 employees at one Exxon Chemical plant, for example, designed a process calling for people to schedule their own coaching sessions with any two other employees. On days and at times selected by the person to be observed, two observers show up at the individual's worksite and use a CBC to conduct a systematic, 30-minute observation session. This plant started with only one scheduled observation per month, and observers were selected from a list of volunteers who had received special coaching training. One year later, employees scheduled two observations per month, and any plant employee could be called on to coach.

The Exxon procedure is markedly different than the "planned 60-second actively caring review" implemented at a Hoechst Celanese plant. For this one-on-one coaching process, all employees attempt to complete a one-minute observation of another employee's work practices in five general categories: body position, personal apparel, housekeeping, tools/equipment, and operating procedures. The initial plant goal was for each of the 800 employees to complete one 60-second behavioral observation every day. Results were entered into a computer file for a behavioral safety analysis of the work culture.

The CBC used for the one-minute coaching observations is shown in Figure 11.5. The front of each card includes the five behavioral categories, a column to check "safe" or "at risk" per category, and columns ("feedback targets") to write comments about the observations. These comments facilitate a feedback session following the observation session, if it is convenient. The back of this CBC includes examples ("memory joggers") related to each behavioral category on the front of the card. These examples summarize the category definitions developed by the CBC steering committee, and determine whether "safe" or "at risk" should be checked on the front of the card.

CRITICAL FEATURES OF THE OBSERVATION PROCESS. Duration, frequency, and scheduling procedures of CBC observations vary widely; still, there are a few common features. First and foremost, the observer must ask permission before beginning an observation process. And the name of the person observed must never be recorded. To build trust and increase participation, a "no" to a request to observe must be honored.

Asking permission to observe serves notice to work safely, and thus biases the observation data, right? In other words, when workers give permission to be coached, their attention to safety will likely increase and they will try to follow all safety procedures. It's possible, though, for people to overlook safety precautions, even when trying their best. They could be unconsciously at risk. When people give permission to be coached, their willingness to accept and appreciate feedback, even when it is corrective, is maximized.

What if people sneak around and conduct behavioral observations with no warning? This is indeed an unbiased plant-wide audit of work practices. It might even be accepted if those observed were not identified. However, if

| Observer: | Location: | Date: |

AUDIT CATEGORY	SAFE	AT RISK	FEEDBACK TARGETS	
			SAFE	AT RISK
Position				
Safe Apparel				
Housekeeping				
Tools/Equipment				
Procedures				
Total				

FRONT OF ONE-MINUTE AUDIT CARD

OBSERVATION TARGETS	SAFE	AT RISK	OBSERVATION TARGETS	SAFE	AT RISK
Position * Line of Fire * Falling * Pinch Points * Lifting			**Tools/Equipment** * Condition * Use * Guards		
Safe Apparel * Hair * Clothes * Jewelry * PPE			**Procedures** * SOPs * JSAs * Permits * Lockout * Barricade * Equipment Release		
Housekeeping * Floor * Equipment * Storage of materials					

BACK OF ONE-MINUTE AUDIT CARD

Figure 11.5. The critical behavior checklist (CBC) used for one-minute observations.

one-on-one coaching is added to this procedure, an atmosphere of mistrust can develop.

Safety coaching should not be a way to enforce rules or play "gotcha." It needs to be seen as a process to help people develop safe work habits through supportive and constructive feedback. Giving corrective feedback after "catching" an individual off-guard performing an at-risk behavior will likely lead to defensiveness and lack of appreciation, even for a well-intentioned effort. It can also reduce interpersonal trust and alienate a person toward the entire safety coaching process.

FEEDBACK IS ESSENTIAL. Each observation process with a CBC provides for tallying and graphing results as "percent safe behavior" on a group feedback chart. All checks for safe observations can be added and divided by the total number of checks (safe plus at-risk behaviors). The result is multiplied by 100 to yield "percent safe behavior."

Applying this formula to checks written on the front of the CBC shown in Figure 11.5 results in a conservative estimate of "percent safe behavior." That's because a "safe" check mark on the CBC in this application means that each separate behavior of a certain category is marked "safe" on the back. Thus, this calculation requires all behaviors relating to a particular observation category to be safe for a "safe" designation. A calculation based on individual behaviors rather than on an "all-in-one" classification of a "safe" or "at-risk" category generally results in higher percentages.

There is no best way to do these calculations. What's important is for participants to understand the meaning of the feedback percentages. As shown in the lower half of Figure 11.6, these percentages can be readily displayed on a group feedback chart or graph. While feedback percentages are valuable, it is vital to realize that the process is more important than the numbers.

The true value of the coaching process is not in the behavioral data, which are no doubt biased by uncontrollable factors, but in the behavior-based interaction between employees. I've actually seen observers get so caught up in recording the numbers, such as frequency of safe and at-risk behaviors, that they let coworkers continue to perform an at-risk behavior while they observe and check columns on a CBC.

An individual's safety must come before the numbers in any observation process. When observers see an at-risk behavior that immediately threatens a person's health or safety, they should intervene at once. They can usually pick up the observation process afterwards. On the other hand, if the CBC was partially completed before they stepped in, it might be most convenient to communicate other observations, especially if there are some safe behaviors to report. This "deposit" will help compensate for the "withdrawal" that was probably implicated by the need to stop a risky behavior at once.

Figure 11.6. Feedback from a critical behavior checklist can be given one-on-one and in groups.

"A" FOR ANALYZE

When interpreting observations, safety coaches draw on their understanding of the three-term contingency or ABC principle (for Activator, Behavior, Consequence) discussed earlier in Chapter 8. They realize observable reasons usually exist for why safe or risky behaviors occur. They know certain dangerous behaviors are triggered by activators such as work demands, risky example-setting by peers, and inconsistent messages from management. They also appreciate the fact that at-risk behaviors are often motivated by one or more natural consequences, including comfort, convenience, work breaks, and approval from peers or work supervisors.

This understanding is critical if safety coaching is to be a "fact finding" rather than "fault finding" process. And it leads to an objective and constructive analysis of the situations observed. This is how people discover the reasons behind at-risk behaviors, and design interventions to decrease them.

An ABC analysis can be done before giving feedback to the person observed, or during the one-on-one feedback process. Discussing the activators and consequences that possibly influenced certain work practices can lead to environment or system improvements for decreasing at-risk behavior.

- Was the behavior observed activated by a work demand or a desire to go on break or leave work early?
- Does the design of the equipment or environment, or the ergonomic design of the task, influence at-risk behavior?
- Is certain personal protective equipment uncomfortable or difficult to use?
- Are fellow workers or supervisors activating dangerous behavior by requesting or demanding an excessive work pace?
- Are certain people motivating at-risk behavior from others by giving rewarding consequences, like words of appreciation, for work done quickly at the expense of safety?

Answers to these and other questions are explored in the next phase of safety coaching—the heart of the process.

"C" FOR COMMUNICATE

A good coach is a good communicator. This means being an "active" listener and persuasive speaker. Since none of us are born with these skills, communication training sessions that incorporate role-playing exercises can be invaluable in developing the confidence and competence needed to send and receive behavioral feedback. Such training should emphasize the need to separate behavior (actions) from person factors (attitudes and feelings). This enables corrective feedback without "stepping on" feelings.

People need to understand that anyone can be at risk without even realizing it, as in "unconscious incompetence," and performance can only improve with behavior-specific feedback. Once this fact is established, corrective feedback that is appropriately given will be appreciated, regardless of who's giving the feedback. Work status is not a factor.

INITIATING FEEDBACK. I remember key aspects of effective verbal presentation with the **SOFTEN** acronym listed in Figure 11.7. First, it's important for the observer to initiate communication with a friendly smile and an open (flexible) perspective. "Territory" reflects the need to respect the fact that you are encroaching on another person's work area. You should ask the person where it would be appropriate and safe to talk. It's also important to maintain a proper physical distance during this interaction.

Standing too close to another person can cause interference and discomfort. There are prominent cultural differences in interpersonal distance norms. In the United States, communicating closer than 18 inches with another person—measured nose to nose—is considered an intimate distance, with 0 to 6 inches reserved for comforting, protesting, lovemaking, wrestling, and other full-contact behaviors. The far phase of the intimate zone (6 to 18 inches) is used by individuals who are on very close terms.

Safety coaches in the United States should most likely communicate at a "personal distance" (18 inches to 4 feet). According to research by Edward Hall (1966), the near phase of the personal distance (18 to 30 inches) is reserved for those who are familiar with one another and on good terms. The

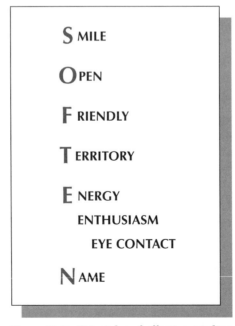

S MILE

O PEN

F RIENDLY

T ERRITORY

E NERGY

ENTHUSIASM

EYE CONTACT

N AME

Figure 11.7. Principles of effective sending
can be remembered with SOFTEN.

far phase of the personal zone (2.5 to 4 feet) is typically used for social interactions between friends and acquaintances. This is likely to be the most common interaction "zone" for a workplace coaching session. Some coaching communication might occur at the near phase (4 to 7 feet) of Hall's "social distance," which is typical for unacquainted individuals interacting informally. These distance recommendations are not hard and fast rules of conduct, but rather personal territory norms we need to consider .

Figure 11.8. Body language communicates more than words.

The "E" of our acronym represents three important directives to remember when coaching—energy, enthusiasm, and eye contact. Your energy and enthusiasm can activate concern and caring on the part of the person you're communicating with. We all know that excited, committed coaches can make "true believers" out of the troops—and that indifferent or distracted coaches can have the opposite effect. Actually, as depicted in Figure 11.8, our body language speaks louder than words.

Finally, we need to remember that the dearest word to someone's ears is his or her own name. Refer to the other person by name, but make it clear the behavioral observations you have recorded will remain anonymous.

INDIVIDUAL FEEDBACK. Whether the aim is to support or correct behavior, feedback should be specific and timely. It should specify a particular behavior, and occur soon after the target behavior is performed. Also, it should be private, given one-on-one to avoid any interference or embarrassment from others. Corrective feedback is most effective if the alternative safe behavior is specified and potential solutions for eliminating the at-risk behavior are discussed.

Anyone giving feedback must actively listen to reactions. This is how a safety coach shows sincere concern for the feelings and self-esteem of the person on the receiving end of feedback. The best listeners give empathic attention with facial cues and posture, paraphrase to check understanding, prompt for more details ("Tell me more"), accept stated feelings without interpretation, and avoid arrogance (such as, "When I worked in your department, I always worked safely").

Figure 11.9 reviews the critical characteristics of effective rewarding (or supportive) and correcting feedback. This figure can be used as a guide for group practice sessions. Because it's not easy to give safety feedback prop-

USE REWARDING FEEDBACK TO SUPPORT SAFE BEHAVIOR
- Give the feedback one-on-one and privately
- Give the feedback as soon as possible after the observation process
- Identify the safe behavior(s) observed
- Be sincere and genuine
- Express personal appreciation for setting the right example for others

Use Correcting Feedback to Decrease At-Risk Behavior
- Give the feedback one-on-one and privately
- Give the feedback as soon as possible after the observation process
- Begin with acknowledgment of safe behavior(s) observed
- Identify the at-risk behavior(s) observed
- Specify the safe alternative to the at-risk behavior(s)
- Indicate concern for the person's welfare
- Request commitment to avoid the at-risk behavior(s)
- Thank the individual for commitment to continuous improvement

Figure 11.9. Key characteristics of rewarding and correcting feedback.

erly and because many people feel awkward or uncomfortable doing it, practice is important.

"H" FOR HELP

The word *help* summarizes what a safety coach is all about. The purpose is to help an individual prevent injury by supporting safe work practices and correcting at-risk practices. It's critical, of course, that a coach's help is accepted. The four letters in the word **help** outline strategies to increase the probability that a coach's advice, directions, or feedback will be appreciated.

HUMOR. Safety is certainly a serious matter, but sometimes a little humor can add spice to our communications, increasing interest and acceptance. It can take the sting out of what some find to be an awkward situation. In fact, researchers have shown that laughter can reduce distress and even benefit our immune systems.

ESTEEM. People who feel inadequate, unappreciated, or unimportant are not as likely to go beyond "the call of duty" to benefit the safety of themselves or others as are people who feel capable and valuable (see Chapter 13 for support of this argument).

The most effective coaches choose their words carefully, emphasizing the positive over the negative, to build or avoid lessening another person's self-esteem. Although Figure 11.10 is humorous, it is unfortunately an accurate

Figure 11.10. Feedback communications more often depreciate than appreciate a person's self-esteem.

portrayal of the atmosphere in many organizational cultures, including the university environment in which I've worked for almost 30 years.

LISTEN. One of the most powerful and convenient ways to build self-esteem is to listen attentively to another person. This sends the signal that the listener cares about the person and his or her situation. And it builds self-esteem: "I must be valuable to the organization because my opinion is appreciated." After a safety coach listens actively, his or her message is more likely to be heard and accepted.

PRAISE. Praising others for their specific accomplishments is another powerful way to build self-esteem. And if the praise targets a particular behavior, the probability of the behavior recurring increases. This reflects the basic principle of positive reinforcement, and motivates people to continue their safe work practices and look out for the safety of coworkers.

Behavior-focused praising is a powerful rewarding consequence which not only increases the behavior it follows, but also increases a person's self-esteem. This in turn increases the individual's willingness to actively care for the safety of others, as I discuss more completely in Part Four of this text.

In Conclusion

Safety coaching is a key intervention process for developing and maintaining a Total Safety Culture. In fact, the more employees who effectively apply the principles of safety coaching discussed here, the closer an organization will come to achieving a Total Safety Culture. The same is true for preventing injury in the community and among our immediate family members at home. Indeed, we need to practice the principles of safety coaching in every situation where an injury could occur following at-risk behavior.

Systematic safety coaching throughout a work culture is certainly feasible in most settings. Large-scale success requires time and resources to develop materials, train necessary personnel, establish support mechanisms, monitor progress, and continually improve the process and support mechanisms whenever possible. For example, the following questions need to be answered at the start of developing an initial action plan:

- Who will develop the critical behavior checklist (CBC)?
- How extensive will the first CBC be?
- What information will be used to define critical behaviors?
- How will safety coaches be trained and receive practice and feedback?
- How many coaches will be trained initially, and how can additional people volunteer to participate as safety coaches?
- How will the coaching sessions be scheduled, how often will people be coached, and how long will the coaching sessions last?
- Where will the group feedback graphs be posted, and who will be responsible for preparing the displays of safe behavior percentages?
- Who will be on the steering committee to oversee the safety coaching process, answer these and other questions about process implementation, maintain records, monitor progress, and refine procedural components whenever necessary?

This doesn't cover all the issues, yet the list might appear overwhelming at first. Organizational cultures vary widely according to personnel, history, policy, the work process, environmental factors, and current contingencies. So implementation procedures need to be customized. There must be significant input from the people protected by the coaching process and from whom long-term participation is needed. It's likely to take significant time, effort, and resources to achieve a plant-wide safety coaching process. With this end in mind, I recommend starting small to build confidence and optimism on small-win accomplishments. Then with patience and diligence, set long-term goals for continuous improvement. Remember to celebrate achievements that reflect successive approximations of your vision—an organization of people who consistently coach each other effectively to increase safe work practices and decrease at-risk behaviors.

Part Four

Actively Caring for Safety

Chapter 12

Understanding Actively Caring

Actively caring is planned and purposeful behavior, directed at environment, person, or behavior factors. It is reactive or proactive, and direct or indirect. Direct, proactive, and behavior-focused caring is most challenging, but in many situations it is most important for large-scale injury prevention. This chapter discusses conditions and situations that inhibit actively caring behavior. We need to understand why people resist opportunities to actively care for safety. Then we can develop interventions to increase this desired behavior, which is critical for achieving a Total Safety Culture.

We cannot live only for ourselves. A thousand fibers connect us with our fellow men; and among those fibers, as sympathetic threads, our actions run as causes; and they come back to us as effects.

This quotation from Herman Melville appeared in a popular paperback entitled *Random Acts of Kindness* (p. 31). Here the editors of Conari Press (1993) introduced the idea of randomly showing kindness or generosity toward others for no ulterior motive except to benefit humanity. This notion seems quite analogous to the actively caring concept I've discussed earlier in various contexts. Indeed, a recurring theme in this book is that a Total Safety Culture can only be achieved if people intervene regularly to protect and promote the safety and health of others. But actively caring is not usually random. It is planned and purposeful. Plus, as implied in Melville's quote, actively caring behaviors (actions) are supported by positive consequences (effects). Sometimes the consequences are immediate, as when someone expresses their appreciation for an act of caring. Or they are delayed but powerful, as in working with care to develop a safer work setting and prevent injuries.

Part Four of this text addresses the need to increase actively caring behavior throughout a culture, and to get the maximum safety and health benefits from this type of behavioral intervention. Psychologists have identified conditions and person states that influence people's willingness to actively care for the safety or health of others. I'll present these and link them to practical things we can do to increase the occurrence of active caring.

What is Actively Caring?

Figure 12.1 presents a simple flow chart summarizing the basic approach to culture change presented up to this point. We start a culture-change mission with a vision or ultimate purpose—for example, to achieve a Total Safety Culture. With group consensus supporting the vision, we develop procedures or action plans to accomplish our mission. These are reflected in process-oriented goals which hopefully activate goal-related behaviors.

Appropriate goal-setting, as I described in Chapter 9, self-affirmations, and a positive attitude can activate behaviors to achieve goals and visions. But we must not forget one of B. F. Skinner's most important legacies—"selection by consequences" (Skinner, 1981). As depicted in Figure 12.1, consequences are needed to support the right behaviors and correct wrong ones. Without support for the "right stuff," good intentions and initial efforts can fade away. Sometimes natural consequences are available to motivate desired behaviors, but often—especially in safety—consequence-contingencies need to be managed to motivate the behavior needed to achieve our goals.

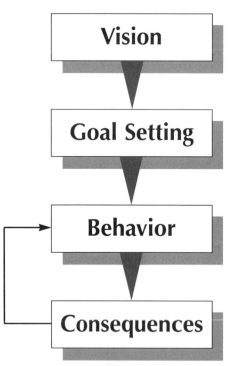

Figure 12.1. A Total Safety Culture requires vision and behavior management.

In Figure 12.2 a new box is added to the basic flow diagram in Figure 12.1. My point is simple but extremely important: Vision, goals and consequence-contingencies are not sufficient for culture change. People need to actively care about goals, action plans, and consequences. They need to believe in and own the vision. They need to feel obligated to work toward attaining goals that support the vision. And they need to give rewarding or corrective feedback to increase behaviors consistent with vision-relevant goals. This is the key to continuous improvement and achieving a Total Safety Culture.

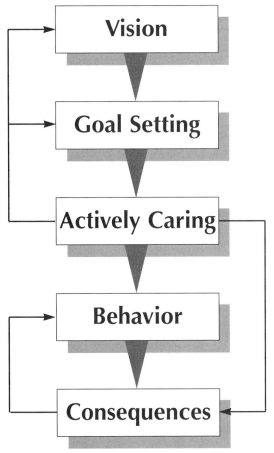

Figure 12.2. Continuous improvement requires actively caring.

THREE WAYS TO ACTIVELY CARE

The Safety Triad introduced in Chapter 3 is useful to categorize actively caring behaviors. These behaviors can address environment factors, person factors, or behaviors. When people alter environmental conditions, or reorganize or redistribute resources in an attempt to benefit others, they are actively caring from an environment perspective. Actively caring safety behaviors in this category include: attending to housekeeping details, posting a warning sign near an environmental hazard, designing a guard for a machine, locking out the energy source to production equipment, and cleaning up a spill.

Person-based active caring occurs when we attempt to make other people feel better. We address their emotions, attitudes or mood states. Actively listening to others, inquiring with concern about another person's difficulties, complimenting an individual's personal appearance, and sending a get-well card are examples. This type of active caring is likely to boost a person's self-esteem, optimism, or belongingness—which in turn increases their propensity to actively care. I'll discuss this in detail in Chapter 13. Also included here are reactive behaviors performed in crisis situations. For example, if you pull someone out of an equipment pinch point or administer cardiopulmonary resuscitation, you're actively caring from a person-based perspective.

Behavior-focused active caring is most useful, and most challenging, from a proactive perspective. This happens when you attempt to direct another person's safe behavior. When we teach others about safe work practices or provide rewarding or corrective feedback regarding observed behavior, we are actively caring from a behavior focus. Obviously, the one-on-one coaching process described in Chapter 11 represents behavior-focused actively caring. Figure 12.3 depicts the three basic ways to actively care.

153

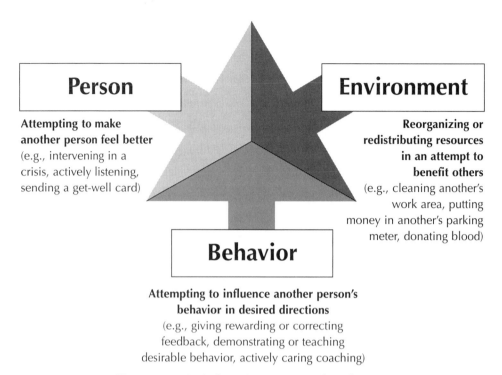

Person

**Attempting to make
another person feel better**
(e.g., intervening in a
crisis, actively listening,
sending a get-well card)

Environment

**Reorganizing or
redistributing resources
in an attempt to
benefit others**
(e.g., cleaning another's
work area, putting
money in another's parking
meter, donating blood)

Behavior

**Attempting to influence another person's
behavior in desired directions**
(e.g., giving rewarding or correcting
feedback, demonstrating or teaching
desirable behavior, actively caring coaching)

Figure 12.3. Actively caring can target three factors.

WHY CATEGORIZE ACTIVELY CARING BEHAVIORS?

So why go to the trouble of categorizing actively caring behaviors? Good question! I think it's useful to consider what these behaviors are trying to accomplish, and realize the relative difficulty in performing each of them. Environment-focused active caring might be easiest for some people because it usually does not involve interpersonal interaction. When people contribute to a charity, donate blood, or complete a organ donor card, they do not interact personally with the recipient of the contribution. These behaviors are certainly commendable and may represent significant commitment and effort, but the absence of personal encounters between giver and receiver warrants separate study from other types of actively caring behavior.

Certain conditions and person characterisitics might facilitate or inhibit one type of actively caring behavior and not the other. For example, communication skills are needed to actively care on the personal or behavioral level. And different skills usually come into play. Behavior-focused active caring is more direct and usually more intrusive than person-focused active caring. It's more risky and potentially confrontational to attempt to direct or motivate another person's behavior than it is to demonstrate concern, respect or empathy for someone.

Classifying actively caring behaviors also provides insight into their benefits and liabilities. Both Brown (1991) and the editors of Conari Press (1993) recommend we feed expired parking meters to keep people from paying excessive fines. Let's consider the behavioral impact of this environment-focused "random act of kindness." What will the vehicle owner think when finding an unexpired parking meter? Could this lead to a belief that parking meters are unreliable—and to further mismanagement of time? Is there a price to pay in people becoming less responsible about sharing public parking spaces?

Considering the long-term and large-scale impact of some actively caring strategies might suggest other approaches. In the parking meter situation, for example, the potential impact would be improved by adding some behavior-focused active caring. Along with feeding the expired parking meter, why not place a note under the vehicle's windshield wiper explaining the act? The note might also include some time management hints. This additional step might not only improve behavior, but set an example. The recipient of the note is probably more inclined to actively care for someone else.

Each type of actively caring behavior can be direct or indirect, with direct behavior requiring effective communication strategies. For instance, leaving a note to explain an actively caring act does not involve interpersonal confrontation. Similarly, you can report an individual's safe or at-risk behavior to a supervisor, and eliminate the need for special communication skills. In the same vein, person-focused actively caring does not always involve interpersonal dialogue. You can send someone a get-well card or leave a friendly uplifting statement on an answering machine or e-mail. It's also possible that environment-focused acts will include personal confrontation, say if you deliver a contribution to a needy individual. This additional category for actively caring behavior is illustrated in Figure 12.4.

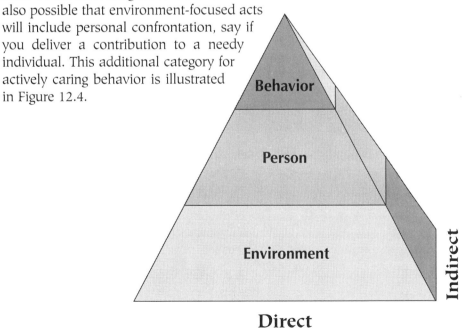

Figure 12.4. Actively caring is usually most challenging and useful when direct and behavior-focused.

AN ILLUSTRATIVE ANECDOTE

Several years ago I was driving on a toll road in Norfolk, Virginia, en route to the Fort Eustis Army Base, where a transportation safety conference was being held. Three of my students were with me, each scheduled to give a 15-minute talk at the conference. This was to be their first professional presentation, and they appeared quite distressed. Each was paging frantically through their notes making last-minute adjustments.

"Were you this nervous, Doc, when you gave your first professional address?" one student asked.

"No, I don't think so," I replied in jest, "I was obviously better prepared."

"Can we just read our paper?" asked another.

"Absolutely not." I retorted, "Anyone could read your paper. It's much more professional and instructive to just talk about your paper informally with the audience."

Naturally, this conversation just caused more anxiety and distress for my students. Something had to be done to distract them, to break the tension. As we approached the first of several toll booths along the highway, I thought of an actively caring solution. After paying a quarter for my vehicle, I handed the attendant another quarter and said, "This is for the vehicle behind us, the driver is using a safety belt and deserves the recognition." My students put down their papers and watched the attendant explain to the driver that we paid her toll because she was buckled up. Because we slowed down to observe this, the driver caught up with us, pulled next to us in the right lane, and acknowledged our actively caring behavior with a "shoulder belt salute"—a smile and tug on her shoulder strap.

At the next toll booth, the driver of the vehicle behind us was not buckled up, but that didn't stop me. I gave the attendant an extra quarter and said, "This is for the vehicle behind us, please ask the driver to buckle up." Again we slowed down to watch, and to our delight the driver buckled up on the spot. When the vehicle passed us, the driver gave us a smile and a "thumbs up" sign.

I kept doing this at every toll booth until exiting the highway, by which time my students had almost forgotten about their papers. They seemed relaxed and at ease when entering the conference room, and each gave an excellent presentation. Later we discussed how the toll booth intervention actively took their minds off their papers and their anxiety.

Brown (1991) recommended that his son occasionally pay the toll for the vehicle behind him. This is redistribution of resources. It's also actively caring with an environment focus. By adding a safety-belt message, I was able to accomplish more than the "random act of kindness" suggested by Brown (1991) and the editors of Conari Press (1993). I was able to support and reward those who were already buckled up, and influence other drivers to buckle up. In other words, realizing the special value of behavior-based actively caring enabled me to get more benefit from an environment-focused strategy with very little extra effort. This behavior-based effort was particularly convenient and effortless because it was indirect. You can see how the system for categorizing actively caring behavior allows us to analyze real and potential acts, and then consider ways to increase their impact.

A HIERARCHY OF NEEDS

The hierarchy of needs proposed by the humanist Abraham Maslow (1943) is probably the most popular theory of motivation. It's taught in a variety of college courses, including introductory classes in psychology, sociology, economics, marketing, human factors, and systems management. It's considered a stage theory: categories of needs are arranged hierarchically, and we don't attempt to satisfy needs at one stage or level until the needs at the lower stages are satisfied.

First, we are motivated to fulfill our physiological needs which include basic survival requirements for food, water, shelter and sleep. After these needs are under control, we are motivated by safety and security needs—the desire to feel secure and protected from future dangers. When we prepare for future physiological needs, we are proactively working to satisfy our need for safety and security.

The next motivational stage includes our social acceptance needs—the need to have friends and to feel like we belong. When these needs are gratified, our concern focuses on self-esteem, the desire to develop self-respect, gain the approval of others, and achieve success.

When I ask audiences to tell me the highest level of Maslow's Hierarchy of Needs, several people usually shout "self-actualization." When I ask for the meaning of "self-actualization," however, I receive limited or no reaction. This is probably because the concept of being self-actualized is rather vague and ambiguous. In general terms, we reach a level of self-actualization when we believe we have become the best we can be, taking the fullest advantage of our potential as human beings. We are working to reach this level when we strive to be as productive and creative as possible. Once accomplished, we possess a feeling of brotherhood and affection for all human beings, and a desire to help humanity as members of a single family—the human race. Perhaps it's fair to say that these individuals are ready to actively care.

Maslow's Hierarchy of Needs is illustrated in Figure 12.5, but self-actualization is not at the top. Maslow (1971) revised his renowned hierarchy shortly before his death in 1970 to put self-transcendence above self-actualization. Transcending the self means going beyond self-interest and is quite analogous to the actively caring concept. According to Viktor Frankl (1962), for exam-

Figure 12.5. The highest need in Maslow's hierarchy reflects actively caring.

ple, self-transcendence includes giving ourselves to a cause or another person, and is the ultimate state of existence for the healthy person. Thus, after satisfying needs for self-preservation, safety and security, acceptance, self-esteem, and self-actualization, people can be motivated to reach the ultimate state of self-transcendence by reaching out to help others—to actively care.

The Psychology of Actively Caring

Walking home on March 13, 1964, Catherine (Kitty) Genovese reached her apartment in Queens, New York, at 3:30 a.m. Suddenly, a man approached with a knife, stabbed her repeatedly, and then raped her.

When Kitty screamed "Oh my God, he stabbed me! Please help me!" into the early morning stillness, lights went on and windows opened in nearby buildings. Seeing the lights, the attacker fled; but when he saw no one come to the victim's aid, he returned to stab her eight more times and rape her again. The murder and rape lasted more than 30 minutes, and was witnessed by 38 neighbors. One couple pulled up chairs to their window and turned off the lights so they could get a better view. Only after the murderer and rapist departed for good did anyone phone the police. When the neighbors were questioned about their lack of intervention, they couldn't explain it.

LESSONS FROM RESEARCH

Professors Bibb Latané, John Darley and colleagues (1970) studied bystander apathy by staging emergency events observed by varying numbers of individuals. Then they systematically recorded the speed at which one or more persons came to the victim's rescue. In the most controlled experiments, the observers sat in separate cubicles (as depicted in Figure 12.6) and could not be influenced by the body language of other subjects.

In the first study of this type, the subjects introduced themselves and discussed problems associated with living in an urban environment. In each condition, the first individual introduced himself and then casually mentioned he had epilepsy and that the pressures of city life made him prone to seizures. During the course of the discussion over the intercom, he became increasingly loud and incoherent, choking, gasping, and crying out before lapsing into silence. The experimenters measured how quickly the subjects left their cubicles to help him.

When a subject believed that he or she was the sole witness, 85 percent left their cubicles within three minutes to intervene. But only 62 percent of the subjects who believed one other witness was present left their cubicle to intervene, and only 31 percent of those who thought four other witnesses were available attempted to intervene. Within three to six minutes after the seizure began, 100 percent of the lone subjects, 81 percent of the subjects with one presumed witness, and 62 percent of the subjects with four other bystanders left their cubicles to intervene.

The reduced tendency of observers of an emergency to help a victim when they believe other potential helpers are available has

Figure 12.6. Subjects in the Latané and Darley experiment could not see each other and thought they were conversing with one, two, or five other individuals.

been termed the *bystander effect* and has been replicated in several situations. Researchers have systematically explored reasons for the bystander effect and have identified conditions influencing this phenomenon. The results most relevant to safety management are reviewed here.

DIFFUSION OF RESPONSIBILITY. It's likely many observers of the Kitty Genovese rape and murder assumed that another witness would call the police, or attempt to scare away the assailant. Perhaps some observers waited for a witness more capable than they to rescue Kitty. Does this factor contribute to lack of intervention for occupational safety? Do people overlook environmental hazards or at-risk behaviors in the workplace because they presume someone else will make the correction? Perhaps some people assume, "If the employees who work in the work area don't care enough to remove the hazard or correct the risk, why should I?"

A HELPING NORM. Many, if not most, U.S. citizens are raised to be independent rather than interdependent. But intervening for the benefit of others, whether reactively in a crisis situation or proactively to prevent a crisis, requires sincere belief and commitment toward interdependence. Social psychologists refer to a "social responsibility norm" as the belief that people should help those who need help. Subjects who scored high on a measure of this norm, as a result of upbringing during childhood or special training sessions, were more likely to intervene in a bystander intervention study, regardless of the number of other witnesses.

KNOWING WHAT TO DO. When people know what to do in a crisis, they do not fear making a fool out of themselves and do not wait for another, more appropriate person to intervene. The bystander effect was eliminated when observers had certain competencies, such as training in first-aid treatment, which enabled them to take charge of the situation. In other words, when observers believed they had the appropriate tools to help, bystander apathy was decreased or eliminated.

This conclusion is also relevant for proactive or preventive action, as in safety intervention. When people receive tools to improve safety, and believe the tools will be accepted and effective to prevent injuries, bystander apathy for safety will decrease. This implies, of course, the need to (a) promote a social responsibility or interdependence norm throughout the culture, and (b) teach and support specific intervention strategies or tools to prevent workplace injuries.

IT'S IMPORTANT TO BELONG. Researchers demonstrated reduced bystander apathy when observers knew one another and had developed a sense of belonging or mutual respect from prior interactions. Most, if not all, of the witnesses to Kitty Genovese's murder did not know her personally, and it's likely the neighbors did not feel a sense of comradeship or community with one another. Situations and interactions that reduce a "we-they" or territorial perspective and increase feelings of togetherness or community will increase the likelihood of people looking out for each other.

MOOD STATES. Several social psychology studies have found that people are more likely to offer help when they are in a good mood. And the mood states that facilitated helping behavior were created very easily—by arranging for potential helpers to find a dime in a phone booth, giving them a cookie, showing them a comedy film, or providing pleasant aromas. Are these findings relevant for occupational safety?

Daily events can elevate or depress our moods. Some events are controllable, some are not. Clearly, the nature of our interactions with others can have a dramatic impact on the mood of everyone involved. Perhaps remembering the research on mood and its effect will motivate us to adjust our interactive behaviors with coworkers. We should also interact in a way that could influence a person's beliefs or expectations in certain directions, as explained next.

Deciding to Actively Care

As a result of their seminal research, Latané and Darley (1968, 1970) proposed that an observer makes five sequential decisions before helping a victim. The five decisions (depicted in Figure 12.7) are influenced by the situation or environmental context in which the emergency occurs, the nature of the emergency, the presence of other bystanders and their reactions, and relevant social norms and rules. While the model was developed to evaluate intervention in emergency situations—where there is need for direct, reactive, person-focused actively caring—it is quite relevant for the other types of actively caring.

STEP 1: IS SOMETHING WRONG?

The first step in deciding whether to intervene is simply noticing that something is wrong. Some situations or events naturally attract more attention than others. This point relates to the discussion in Chapter 9 about relative attention and habituation to various activators. Most emergencies are novel and upset the normal flow of events. However, the onset of an emergency such as a person slipping on a spill or falling down a flight of stairs will attract more attention and helping behavior than the aftermath of an incident, as when a victim is regaining consciousness or rubbing an ankle after a fall. Of course, we should expect much less attention to a nonemergency situation.

Now, even if the need for proactive intervention is noticed, actively caring behavior will not necessarily occur. The observer must interpret the situation as requiring intervention. Which leads us to the next question that must be answered before deciding to intervene.

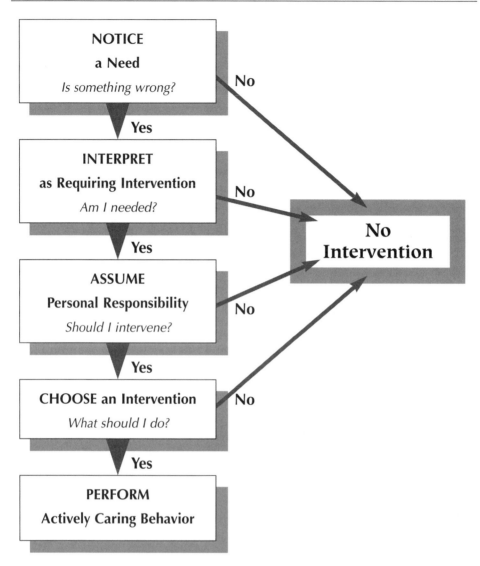

Figure 12.7. Actively caring requires five sequential decisions.

STEP 2: AM I NEEDED?

As shown in Figure 12.8, people can come up with a variety of excuses for not helping. Distress cues, such as cries for help, and the actions of other observers can clarify an event as an emergency. When people are confused, they look to other people for information and guidance. In other words, through observational learning (Chapter 7), people figure out how to interpret an ambiguous event and how to react to it. Thus, the behavior of others is especially important when stimulus cues are not present to clarify a situation as requiring intervention.

Figure 12.8. People give a variety of excuses for not helping.

STEP 3: SHOULD I INTERVENE?

In this stage you ask yourself, "Is it my responsibility to intervene?" The answer would be obvious if you were the only witness to a situation you perceive as an emergency. But you might not answer "yes" to this question when you know that other people are also observing the emergency, or the safety hazard. In this case, you have reason to believe that someone else will intervene, perhaps a person more capable than you. This perception relieves you of personal responsibility. But what happens when everyone believes the other guy will take care of it? This is likely what happened in the Kitty Genovese tragedy.

The challenge in achieving a Total Safety Culture is to convince everyone they have a responsibility to intervene for safety. Indeed, a social norm or expectancy must be established that everyone shares equally in the responsibility to keep everyone safe and healthy. Furthermore, safety leaders or captains need to accept the special responsibility of teaching others any techniques they learn at conferences or group meetings that could increase a person's perceived competence to intervene effectively. All this is easier said than done, of course. Unfortunately, if we don't meet this challenge many people are apt to decide that actively caring safety intervention is not for them. They could feel this way even after viewing an obvious at-risk behavior or condition that would benefit from their immediate action.

STEPS 4 AND 5: WHAT SHOULD I DO?

These last two steps of Latané and Darley's decision model point out the importance of education and training. Education gives people the rationale and principles behind a particular intervention approach. It gives people information to design or refine intervention strategies, leading to a sense of ownership for the particular tools they help to develop. Through training, people learn how to translate principles and rules into specific behaviors or intervention strategies. Safety coach training, for example, should include role-playing exercises so people practice certain communication techniques and receive specific feedback regarding their strengths and weaknesses. The bottom line here is that people who have learned how to intervene effectively through relevant education and training are likely to be successful agents of intervention.

SUMMARY OF THE DECISION FRAMEWORK

In this section I have reviewed the decision process model Latané and Darley proposed as a sequence of choices people make before actively caring on behalf of another person. Although developed to understand the bystander effect in emergency situations, this decision framework is certainly relevant for proactive situations and for each type of actively caring behavior defined in this chapter (direct versus indirect, and environment-focused, person-focused or behavior-focused). The model can help us understand why an individual might not actively care for another person's health and safety, and it can be used to guide the development of strategies to increase the frequency of actively caring behaviors. For example, conditions (activators) will increase the probability of this behavior if they increase the likelihood a person will notice and perceive a need for intervention, and assume personal responsibility for helping. Plus, education and training sessions that increase skills and self-confidence to intervene *effectively* will increase the amount of constructive actively caring behavior occurring throughout a culture.

A Consequence Analysis of Actively Caring

When I related the Kitty Genovese case to my family and asked for opinions, I received a unanimous reaction that I could not readily explain with the decision model discussed above. My wife and two daughters proclaimed that most observers didn't help these women because they feared for their own safety. The perpetrator was armed with a knife. The onlookers could certainly see there was an emergency requiring specific assistance from anyone who would take responsibility. According to an interpretation based on our understanding of the power of consequences, people resisted taking responsibility because they perceived that it could mean more trouble—or potential harm—than it was worth. It was safer to assume that someone else more capable would intervene. According to this consequence model, people hesitated to intervene because they perceived more potential costs than benefits, not because they were apathetic or failed to interpret a need to take personal responsibility.

The matrix in Figure 12.9 combines two levels of cost (low versus high) to the potential intervention agent and the victim in order to predict when actively caring behavior will occur. It's most likely (lower left cell of Figure 12.9) when costs for helping are low, for example, convenient and not dangerous; and costs to the victim for not helping are high, as when the victim is seriously injured.

On the other hand, intervention is least likely when the perceived personal costs for intervening are high, for example, effortful and risky; and the apparent costs to the victim for no intervention are low, as when an experienced worker is performing at-risk behavior with no negative consequences.

The Kitty Genovese case fits the lower right cell of Figure 12.9: high perceived cost for both helper and victim. Although the costs for not helping Kitty were extremely high, resulting ultimately in her death, the costs for helping were also high, in fact potentially fatal. This means significant conflict for the person deciding what to do. People can resolve the conflict by helping indirectly, say by telephoning police or an ambulance; or by reinterpreting the situation. This can be done by presuming someone else will intervene—diffusion of responsibility—or perhaps by rationalizing that the person does not deserve help.

A bystander might rationalize, for example, that Ms. Genovese should not have been walking the streets in that neighborhood at 3:30 in the morning. Rationalization reduces the perceived costs for not intervening, and enables the bystander to ignore the situation without excessive shame or guilt. According to this cost-reward interpretation, when bystanders perceive high costs both for intervening and for not intervening in a crisis, they recognize the need for action, hesitate because of perceived personal costs, and then search for an excuse to do nothing.

The upper left quadrant of Figure 12.9 represents situations most analogous to actively caring for injury prevention. Although a simple low-cost intervention might be called for to correct an environmental hazard or an at-risk behavior, there is no immediate emergency and thus no need for immediate action. There is low perceived cost if no action is taken: "We've been working under these conditions for months and no one has been hurt."

Cost to Bystander for Intervening

		Low	High
Cost to Victim for Not Intervening	Low	Depends on person factors	Intervention improbable
	High	Intervention probable	Diffusion of responsibility

Figure 12.9. Costs to bystanders for intervening, and costs to a victim if there is no intervention, determine the probability of intervention. (Adapted from Piliavin et al, 1981)

Intervention in situations represented by this cost quadrant is most difficult to predict. Many factors can influence perceived consequences that are positive and negative, and small changes in these factors can tilt the cost-reward balance in favor of stepping in or standing back.

Through testimonials and constructive discussions, employees can be convinced that the potential cost of not intervening is higher than they initially thought. This can occur, for example, by considering the large degree of plant-wide exposure to a certain uncorrected hazard. Also, it might be worthwhile to remind people of the large-scale detrimental learning that could occur from the continuous performance of risky behavior. Furthermore, education and role-playing exercises can reduce the perceived personal costs of actively caring. It's also true that person factors, such as mood states discussed earlier, determine whether intervention occurs.

In Conclusion

Actively caring behavior is planned and purposeful. It can be direct or indirect. And its focus is environment, person, or behavior. Active caring that addresses the environment is usually easiest to perform because it doesn't involve interpersonal confrontation. Behavior-focused actively caring is often most proactive, but is most difficult to carry out effectively because it attempts to influence another person's behavior in a nonemergency situation.

The finding that people often refuse to act in a crisis, especially when they can share the responsibility of intervening with others, is quite analogous to most work settings. Hence, it's important to understand the factors that can influence this resistance, referred to as "bystander apathy." For example, people with a sense of social responsibility and comradeship for others at work, and who believe they have personal control in a just world, are more apt to intervene for the safety of others. It's possible to increase these person characteristics among people through policy, procedures, and personal interaction. Increasing these states, and thus the willingness to actively care for safety, is key to achieving a Total Safety Culture and is addressed in the next two chapters.

Chapter 13

The Person-Based Approach to Actively Caring

Our willingness to actively care for others is affected by certain feelings and states of mind. If we have a strong sense of self-esteem, self-efficacy, personal control, optimism, and belongingness, there's a greater chance we'll go beyond the call of duty to help others. Each of these person states are explained in this chapter, and the research supporting direct relationships between these states and actively caring behavior is reviewed. Understanding these connections enables us to design conditions and interventions to increase actively caring behavior throughout an organization or community.

 Our deeds determine us, as much as we determine our deeds.

This quotation from George Eliot indicates that our behaviors influence something about us and implies that good deeds or actively caring behaviors are good for us. They change something about us, and this in turn affects subsequent behavior. Does this mean our actively caring behaviors influence us to actively care even more? It's a nice thought and seems intuitive, but what does it really mean?

This chapter explores a host of questions arising from the concept reflected in Eliot's words: What is it about us that changes as a result of our good deeds, and will this change lead to more good deeds? Can making people more willing to actively care be influenced in ways other than managing activators and consequences to directly change behavior? In other words, can we change something about people that will make them more willing to actively care for the safety and health of others? If answers to these questions can be turned into practical procedures, we'll know how to increase actively caring behaviors throughout a culture.

At the time I was preparing the first draft of this chapter, a heartwarming story appeared in our local newspaper, *The Roanoke Times and World Report*. The newspaper report of the incident is reprinted in Figure 13.1 and it is clearly opposite to the Kitty Genovese tragedy reviewed in Chapter 12. In this case, two individuals, Tywanii Hairston and John McKee, went out of their way to save the life of a truck driver named Don Arthur, whose truck was blocking traffic because he had blacked out. Several individuals had already driven around his truck without intervening—for a variety of possible reasons discussed in Chapter 12. But two individuals did interrupt their routine to actively care, and the result was a life saved.

Ms. Hairston and Mr. McKee went into action, rather than succumb to bystander apathy. Why? Were there special characteristics of the two heroes? Both individuals did have some life-saving training in the past, so as discussed in Chapter 12, they might have felt more responsible than others because they knew what to do. This chapter examines additional person-based reasons (or internal characteristics of people) that could have con-

---

PAIR BREATHES LIFE INTO DRIVER

Tywanii Hairston was on the way to pay her water bill Tuesday when she pulled up behind a Roanoke city truck at a red light on Campbell Avenue.

The light turned green, but the truck didn't budge. After a moment the driver of another car honked the horn, and still the truck didn't move. Hairston thought it had stalled.

As she slowly drove around the truck, she saw a man slumped over the wheel. A former nursing assistant, Hairston parked her car and jumped out to see what was wrong.

Meanwhile, John McKee, a driver for Alert Towing on his way to a job, also had seen the man and stopped. He radioed his dispatcher to call 911, and with the help of another passerby, pulled the man from the truck and laid him on the street.

Together, Hairston and McKee went into action. Neither had ever administered cardiopulmonary resuscitation, or CPR, before, although both knew the procedure. Hairston started giving mouth-to-mouth, and McKee pumped the man's chest until paramedics arrived. Their swift action quite possibly saved Don Arthur's life. On Thursday, Arthur, 60, was in stable condition at Community Hospital, one of only a few who make it back after venturing so close to death.

Figure 13.1. Actively caring behavior saved a life. (From *The Roanoke* [Virginia] *Times*, Friday, Sept. 8, 1995)

tributed to this success story. Did the two intervening agents have personality traits conducive to helping? Or did recent experiences influence their personality state in some beneficial way? Obviously, we'll never know the answer to these questions, but exploring the possibilities can help us better grasp the person factors contributing to actively caring behavior. And if these factors can be altered systematically, then it's possible to increase people's willingness to actively care for the safety and health of others.

Actively Caring from the Inside

Perhaps you recall earlier discussions in this text about "outside" versus "inside" aspects of people. In Chapter 3, for example, I distinguished between behaviors (outside) versus intentions, attitudes and values (inside), and emphasized that we should start with behaviors. A prime principle of behavior-based psychology is that it's easier, especially for large-scale culture change, to "act a person into safe thinking" than it is to address attitudes and values directly in an attempt to "think a person into safe acting." Another key principle of behavior-based psychology is that the consequences of our behavior influence how we feel about the behavior. Generally, positive consequences lead to good feelings or attitudes; negative consequences lead to bad feelings or attitudes.

Long-term behavior change requires people to change "inside" as well as outside. The promise of a positive consequence or the threat of a negative one can maintain the desired behavior while the response-consequence contingencies are in place. But what happens when they are withdrawn? What happens when people are in situations, like at home, when no one is holding them accountable for their behavior? If people do not believe in the safe way of doing something and do not accept safety as a value or a personal mission, don't count on them to choose the safe way when they have the choice. In addition, if people are not self-motivated to keep themselves safe, don't expect them to actively care for the safety of others.

As discussed in Chapter 5, numerous internal and situational factors influence how we perceive activators and consequences. For example, if we see activators and consequences as nongenuine ploys to control us, our attitude about the situation will be negative. If we believe the external contingencies are genuine attempts to help us do the right thing, our attitude will be more positive. Thus, person or internal dynamics determine how we receive activator and consequence information. This can influence whether environmental events enhance or diminish what we do.

In Chapters 9 and 10, I showed how direct manipulations of activators and consequences can influence large-scale behavior change. Now let's see if changes in internal person factors can benefit behavior change. In particular, how do "inside" factors affect actively caring for safety?

PERSON TRAITS VERSUS STATES

Some person factors are presumed to be *traits,* while others are *states.* Theoretically, traits are relatively permanent characteristics of people, and do not vary appreciably over time or across situations. In contrast, states are characteristics that can change moment-to-moment depending on circumstances and personal interactions. When our goals are thwarted, for example, we can be in a state of frustration. And when experience leads us to believe we have little control over events around us, we can be in a state of apathy or helplessness. These states can influence certain behaviors.

Frustration often provokes aggressive behavior; perceptions of helplessness can inhibit constructive behavior or facilitate inactivity. In contrast, certain life experiences can affect positive person states, such as optimism, personal control, self-confidence and belongingness. This in turn increases constructive behavior, including actively caring.

ACTIVELY CARING STATES

I contend that actively caring characteristics internal to people are states, not traits. Plus, certain conditions—including activators and consequences—can influence these psychological states. These states are illustrated in Figure 13.2, a model my associates and I have used many times to stimulate discussions among employees. We talk of specific situations, operations, or incidents that influence their willingness to actively care for the achievement of a Total Safety Culture. Let's examine these influential states in more detail.

SELF-ESTEEM ("I AM VALUABLE"). How do you feel about yourself? Generally good or generally bad? Your level of self-esteem is determined by the extent to which you generally feel good about yourself. If we don't feel good about ourselves, it's unlikely we'll care about making a difference in the lives of others. The better we feel about ourselves, the more willing we are to actively care for the safety and health of other people.

It's important to maintain a healthy level of self-esteem, and to help others raise their self-esteem. Research shows that people with high self-esteem report fewer negative emotions and less depression than people with low self-esteem. Those with higher self-esteem also handle life's stresses better. Recall the discussion of stress versus distress in Chapter 6. Higher self-esteem turns stress into something positive, rather than negative distress.

Researchers have also found that individuals who score higher on measures of self-esteem are less susceptible to outside influences, more confident of achieving personal goals, and make more favorable impressions on others in social situations. And supporting the actively caring model depicted in Figure 13.2, people with higher self-esteem help others more frequently than those scoring lower on a self-esteem scale.

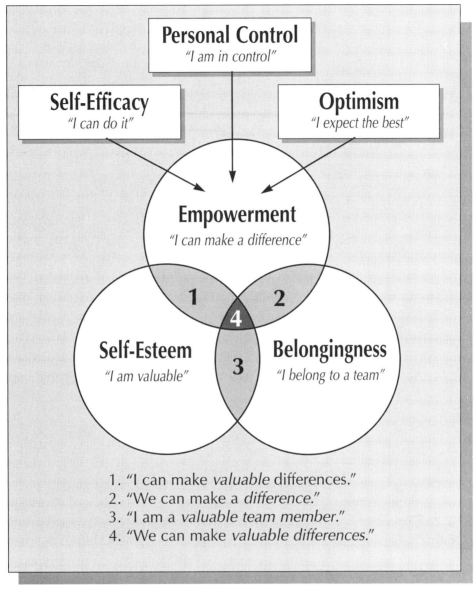

Figure 13.2. Certain person states influence a person's willingness to actively care for the safety and health of others.

Here's something to keep in mind, though. It's also been found that people with high self-esteem are less willing to ask for help than people with low self-esteem. Later in this chapter, I discuss in more detail research that shows a direct relationship between self-esteem and actively caring behavior.

EMPOWERMENT ("I CAN MAKE A DIFFERENCE"). In management literature, empowerment typically refers to delegating authority or responsibility, or sharing decision making. In contrast, the person-based perspective of empowerment focuses on how the person who receives more power or influence reacts. From a psychological perspective, empowerment is a matter of feeling—how does it feel to be more responsible? Can you handle it? This view of empowerment requires the personal belief that "I can make a difference." Perceptions of personal control, self-efficacy, and optimism strengthen this belief. An empowered state is presumed to increase motivation to "make a difference," perhaps by going beyond the call of duty. Let's look more closely at these three factors affecting our sense of ability and personal control—and our propensity to actively care:

Self-efficacy is the idea that "I can do it." This is a key factor in social learning theory, determining whether a therapeutic intervention will succeed over the long term. We're talking about your self-confidence. Dozens of studies have found that subjects who score relatively high on a measure of self-efficacy perform better at a wide range of tasks. They show more commitment to a goal and work harder to pursue it. They demonstrate greater ability and motivation to solve complex problems at work. They have better health and safety habits; and they are more apt to handle stressors positively, rather than with negative distress.

Self-efficacy contributes to self-esteem, and vice versa; but these constructs are different. Simply put, self-esteem refers to a general sense of self-worth; self-efficacy refers to feeling successful or effective at a particular task. Self-efficacy is more focused, and can vary markedly from one task to another. One's level of self-esteem remains rather constant across situations.

When I'm losing to an opponent on the tennis court, my self-efficacy usually drops considerably. However, my self-esteem might not change at all. I might protect my self-esteem by rationalizing that my opponent is younger and more experienced, or that I'm more physically tired and mentally preoccupied than usual. My damaged self-efficacy will undoubtedly lead to reduced optimism about winning the match. If I continue to lose at tennis and run out of excuses, my self-esteem could suffer if I think it's important for me to play tennis well. Now there would be a prominent gap between my real self, a loser at tennis, and ideal self, a winner on the court.

Personal control is the feeling that "I am in control." Professor J. B. Rotter (1966) used the term *locus of control* to refer to a general outlook regarding the location of forces controlling a person's life—internal or external. Those with an internal locus of control believe they usually have direct personal control over significant life events as a result of their knowledge, skill, and abilities. They believe they are captain of their life's ship. In contrast, persons with an external locus of control believe factors like chance, luck or fate play important roles in their lives. In a sense, externals believe they are victims, or sometimes beneficiaries, of circumstances beyond their direct personal control. As depicted in Figure 13.3, however, there are times when everyone likes to feel that their successes resulted from their own efforts.

"COULD I TAKE A MOMENT TO CHANGE MY SHIRT?"

Figure 13.3. There are times when we all want credit for our performance.

Optimism is reflected in the statement, "I expect the best." It's the learned expectation that life events, including personal actions, will turn out well. Compared to pessimists, optimists maintain a sense of humor, perceive problems or challenges in a positive light, and plan for a successful future. *They focus on what they can do rather than on how they feel.* As a result, optimists handle stressors constructively and experience positive stress rather than negative distress. Optimists essentially expect to be successful at whatever they do, and so they work harder than pessimists to reach their goals. As a result, optimists are beneficiaries of the self-fulfilling prophecy.

The self-fulfilling prophecy starts with a personal expectation about one's future performance and ends with that expectation coming true because the individual performs in such a way to make it happen. What do you expect when your boss or supervisor asks to see you? Do you expect the best? Our past experiences with top-down control and the use of negative consequences to influence our behavior often results in pessimistic rather than optimistic expectations. Moreover, our approach to this situation, illustrated in Figure 13.4, can support our negative expectations.

Figure 13.4. When our boss asks to see us, we expect the worst.

If you expect to be punished or reprimanded every time your boss or supervisor calls you into the office, then your body language and demeanor will subtly reflect that expectation. You'll "telegraph" these signals to your boss, who might think, "Scott sure looks guilty, I wonder what he's done that needs to be punished?" However, if you approach the interaction with an optimistic attitude, reflected in your body language and verbal behavior, the results could be more positive. You could, for example, write a different internal script: "No one is perfect, and I might have missed something. Everyone can improve with specific behavioral feedback. If I help to make the interaction constructive, the outcome can only be positive."

It's important to understand that fulfilling a pessimistic prophecy can depreciate our perceptions of personal control, self-efficacy, and even self-esteem. Realizing this should motivate us to do whatever we can to make interpersonal interactions positive and constructive. This will not only increase optimism in a work culture, but also promote a sense of group cohesiveness or belongingness—another person state that facilitates actively caring behavior.

BELONGINGNESS. In his best seller, *The different drum: Community making and peace,* Dr. M. Scott Peck (1979) challenges us to experience a sense of true community with others. We need to develop feelings of belongingness with one another regardless of our political preferences, cultural backgrounds and religious doctrine. We need to transcend our differences, overcome our defenses and prejudices, and develop a deep respect for diversity. He proclaims that we must develop a sense of community or interconnectedness with one another if we are to accomplish our best and ensure our survival as human beings.

It seems intuitive that building a sense of community or belongingness among our coworkers will improve organizational safety. Safety improvement requires interpersonal observation and feedback, and for this to happen, people need to adopt a collective win/win perspective instead of the individualistic win/lose orientation common in many work settings. A sense of belongingness or interconnectedness leads to interpersonal trust and caring—essential features of a Total Safety Culture.

In my numerous group discussions with employees on the belongingness concept, someone inevitably raises the point that a sense of belongingness or community at their plant has decreased over recent years. "We used to be more like family around here" is a common theme. For many companies, growth spurts, continuous turnover—particularly among managers—or "lean and mean" cutbacks have left many employees feeling less connected and trusting. It seems, in some cases, people's need level on Maslow's hierarchy (see Figure 12.5) has regressed from satisfying belongingness needs to concentrating on maintaining job security, in order to keep food on the table.

Figure 13.5 lists a number of special attributes prevalent in most families, where interpersonal trust and belongingness are often optimal. We are willing to actively care in special ways for the members of our immediate family. The result is optimal trust, belongingness, and actively caring behavior for the safety and health of our family members. To the extent we follow

- We use more rewards than penalties with *family* members.
- We don't pick on the mistakes of *family* members.
- We don't rank one *family* member against another.
- We brag on the accomplishments of *family* members.
- We respect the property and personal space of *family* members.
- We pick up after *family* members.
- We correct the at-risk behavior of *family* members.
- We accept the corrective feedback of *family* members.
- We are our brother/sisters keepers of *family* members.
- We actively care because they're *family*.

Figure 13.5. Incorporating an actively caring *family* perspective in an organization will help to cultivate a Total Safety Culture.

the guidelines in Figure 13.5 among members of our "corporate family," we will achieve a Total Safety Culture. Following the principles in Figure 13.5 will develop trust and belongingness among people, and lead to the quantity and quality of actively caring behavior expected among family members—at home and at work.

Measuring Actively Caring States

The Safety Culture Survey which Steve Roberts, Mike Gilmore, and I developed for culture assessment and corporate training programs (Geller *et al*, 1996) includes sub-scales to measure self-esteem, self-efficacy, personal control, optimism, and group cohesion. The twenty questions included in Figure 13.6 were selected from the Actively Caring Person Scale of our Safety Culture Survey. Each of the five actively caring states discussed in this chapter are assessed. There are only four questions per person state, so this should not be considered a reliable nor valid measure of these factors. In other words, don't read too much into this survey. Just respond to each query according to the instructions and then check the answer key in Figure 13.7 to increase your understanding of the five actively caring person states. Comparing the items that measure self-esteem with those that assess self-efficacy, for example, help you understand the distinct difference between these two constructs. Remember, these person factors are presumed to be states that fluctuate from day to day and from situation to situation. The score you get today might be quite different than the one you would obtain on another day under a different set of circumstances.

This is a questionnaire about your beliefs and feelings. Read each statement, then circle the number that best describes your current feelings. There are no "right" or "wrong" answers. This questionnaire only asks about your personal opinions.

	HIGHLY DISAGREE	DISAGREE	NOT SURE	AGREE	HIGHLY AGREE
1. I feel I have a number of good qualities.	1	2	3	4	5
2. Most people I know can do a better job than I can.	1	2	3	4	5
3. On the whole, I am satisfied with myself.	1	2	3	4	5
4. I feel I don't have much to be proud of.	1	2	3	4	5
5. When I make plans, I am certain I can make them work.	1	2	3	4	5
6. I give up on things before completing them.	1	2	3	4	5
7. I avoid challenges.	1	2	3	4	5
8. Failure just makes me try harder.	1	2	3	4	5
9. People who never get injured are just plain lucky.	1	2	3	4	5
10. People's injuries result from their own carelessness.	1	2	3	4	5
11. I am directly responsible for my own safety.	1	2	3	4	5
12. Wishing can make good things happen.	1	2	3	4	5
13. I hardly ever expect things to go my way.	1	2	3	4	5
14. If anything can go wrong for me, it probably will.	1	2	3	4	5
15. I always look on the bright side of things.	1	2	3	4	5
16. I firmly believe that every cloud has a silver lining.	1	2	3	4	5
17. My work group is very close.	1	2	3	4	5
18. I distrust the other workers in my department.	1	2	3	4	5
19. I feel like I really belong to my work group.	1	2	3	4	5
20. I don't understand my coworkers.	1	2	3	4	5

Figure 13.6. Sample items from the Actively Caring Person Scale.

> ► **Self-Esteem** *(items 1-4)* = feelings of self worth and value
> 　　Actual scale = 16 items
> 　　(a) Add numbers for items 1 & 3　　　　　　　　　Total 1 = _____
> 　　(b) Add numbers for items 2 & 4 and subtract from 14　　Total 2 = _____

> ► **Self-Efficacy** *(items 5-8)* = general levels of belief in one's competence
> 　　Actual scale = 23 items
> 　　(a) Add numbers for items 5 & 8　　　　　　　　　Total 1 = _____
> 　　(b) Add numbers for items 6 & 7 and subtract from 14　　Total 2 = _____

> ► **Personal Control** *(items 9-12)* = the extent to which a person believes he or she is responsible for his/her life situation
> 　　Actual scale = 25 items
> 　　(a) Add numbers for items 10 & 11　　　　　　　　Total 1 = _____
> 　　(b) Add numbers for items 9 & 12 and subtract from 14　　Total 2 = _____

> ► **Optimism** *(items 13-16)* = the extent to which a person expects the best will happen for him/her
> 　　Actual scale = 8 items
> 　　(a) Add numbers for items 15 & 16　　　　　　　　Total 1 = _____
> 　　(b) Add numbers for items 13 & 14 and subtract from 14　　Total 2 = _____

> ► **Belongingness** *(items 17-20)* = the perception of group cohesiveness or feelings of togetherness
> 　　Actual scale = 20 items
> 　　(a) Add numbers for items 17 & 19　　　　　　　　Total 1 = _____
> 　　(b) Add numbers for items 16 & 20 and subtract from 14　　Total 2 = _____

> **"ACTIVELY CARING" SCORE** = Sum of Self-Esteem, Self-Efficacy, Optimism, Personal Control, and Belongingness Totals　　　　　　　　**Total Score = _____**

Figure 13.7. Scoring your answers to the twenty person-state items will increase your understanding of the Actively Caring model.

In Conclusion

In this chapter we continued to develop an understanding of actively caring behavior as it relates to injury prevention. A person-based approach was emphasized; we considered subjective factors inside people as potential determinants of active caring. The notion of a general actively caring person-ality was entertained at first, but discarded because of relatively limited empirical support. Even if some people do have permanent traits which make them more or less prone to actively care, it's unlikely we could use this infor-mation to improve safety or prevent injuries. However, we could benefit safe-ty if certain person states influence people's willingness to actively care, and if these states can be manipulated by controllable outside factors.

Five person states were proposed as influencing people's willingness to actively care—self-esteem, self-efficacy, personal control, optimism, and belongingness. Each of these person variables has a prosperous research his-tory in psychology and some of this research relates directly to the actively caring model. Research that tested relationships between person states and actual behavior has generally supported the actively caring model.

A particularly important question is whether actively caring states are both antecedents and consequences of a caring act. It seems intuitive that performing an act of kindness that is effective, accepted, and appreciated could increase a helper's self-esteem, self-efficacy, personal control, opti-mism, and belongingness. This, in turn, should increase the probability of more actively caring behavior. In other words, one act of caring, properly appreciated, should lead to another . . . and another. A self-supporting active-ly caring cycle is likely to occur. You can see how important it is to get active-ly caring behavior started and accepted among a large group of individuals. This challenge is addressed in the next chapter.

Chapter 14

Increasing Actively Caring Behaviors

This chapter integrates principles and procedures from previous chapters to address the most critical question regarding the achievement of a Total Safety Culture. Namely, how can we increase actively caring behavior throughout a work culture? Some conditions and interpersonal techniques facilitate this behavior indirectly by benefiting self-esteem, empowerment, or belongingness. Other procedures can directly boost actively caring with certain activators and consequences. In addition, the social influence principles of reciprocity and consistency can be applied to enhance the caring behavior we desire. In sum, this chapter shows you how to increase the likelihood that people will go beyond their normal routines to help keep people safe.

In a Total Safety Culture . . . people actively care *on a continuous basis for safety.*

This quotation from my 1994 article in *Professional Safety* (Geller, 1994, p.18) reflects both the purpose and fundamental challenge of a Total Safety Culture: Everyone must periodically go beyond his or her personal routine for the safety and health of others. To meet the aim, and the challenge, we need to find ways to increase actively caring behaviors.

So how do we do this at work, in our homes, and the community at large? I hope you can already provide some answers after reading Chapters 12 and 13, where I alluded to some ways to increase actively caring behaviors. Here I want to expand on earlier suggestions and add more. We can classify these approaches as indirect or direct.

Indirect strategies for facilitating actively caring behavior follow from the theory and principles discussed in Chapter 13. The actively caring model supported by research proposes that certain person states inside people increase their willingness to look beyond self-interests and consider the safety of others. Thus, conditions and procedures that increase these states will indirectly increase the probability of active caring. More direct ways to increase these behaviors can be derived from behavior-based principles of learning and social influence. These are discussed in the latter part of this chapter.

Enhancing Actively Caring Person States

Sometimes at seminars and workshops I hear participants express concern that the actively caring person-state model might not be practical. "The concepts are too soft or subjective," is a typical reaction. Employees accept the behavior-based approach because it is straightforward, objective, and clearly applicable to the workplace. But person-based concepts like self-esteem, empowerment, and belongingness appear ambiguous, "touchy-feely," and difficult to deal with. "The concepts sound good and certainly seem important, but how can we get our arms around these 'warm fuzzies' and use them to promote safety?"

These person states are more difficult to define, measure and manage than behaviors. That's why I said early on in this text that it's more cost-effective to work on behaviors first. Whenever the ultimate outcome is behavior change, it's usually most efficient to deal directly with behaviors. However, it's always important to consider people's feelings when designing behavior-change interventions. That's why I recommend against using negative consequences whenever possible. And why I offer ways to design intervention strategies that account for subjective person-states like commitment, ownership, and involvement.

After introducing the actively caring person-state model at my workshops, I often divide participants into discussion groups. I ask group members to define events, situations, or contingencies that decrease and increase the person state assigned to their group. Then I ask the groups to derive simple and feasible action plans to increase their assigned state. This promotes personal and practical understanding of the concept.

Let's take a look at what my workshop participants have come up with for factors and strategies regarding each of these person states:

SELF-ESTEEM

Factors consistently mentioned as shaping self-esteem include communication techniques, reinforcement and punishment contingencies, and leadership styles. Participants suggest a number of ways to build self-esteem, including:

- Provide opportunities for personal learning and peer mentoring
- Increase recognition for desirable behaviors and individual accomplishments
- Solicit and follow up on a person's suggestions

COMMUNICATION STRATEGIES. Figure 14.1 lists eleven words beginning with the letter "A" that imply a specific verbal technique for increasing a person's self-esteem. Each "A" word suggests a slightly different communication approach, from stating simple words of agreement, admiration, appreciation, and approval to acknowledging the achievement and individual creativity of others through active listening and praise. It's also a good idea to argue less and avoid criticizing. Arguments waste time and usually promote a win/lose perspective, and criticism always does more harm than good.

No one likes to be criticized, but some people are more resilient to negative feedback than others. People with a high and stable self-esteem have developed mechanisms to protect themselves. Hence, these people take criticism in stride, protecting their self-esteem, for example, by just denying a mistake, rationalizing that the flack was biased and unwarranted, or optimistically focusing on positive aspects of the verbal exchange. On the other hand, some people are overly sensitive to critical comments. These folks usually have a relatively low or unstable self-esteem, and perceived daggers can devastate them.

When offering corrective feedback, it's critical to be a patient, active listener. Allow the person to make excuses, and don't argue about these. Resist

- **Accept:** appreciate diversity
- **Actively listen:** with verbal & nonverbal behavior
- **Agree:** with verbal & nonverbal behavior
- **Admire:** "attractive dress," "nice tie"
- **Appreciate:** "please " (activator) & "thank you" (consequence)
- **Acknowledge:** the achievements of others
- **Approve:** praise for good behavior
- **Ask:** for feedback, advice, opinions
- **Attend:** lend a helping hand
- **Avoid criticizing:** it won't be accepted anyway
- **Argue less:** arguments are win/lose situations

Figure 14.1. Apply "A" strategies to increase others' self-esteem.

the temptation. Giving excuses is just a way to protect self-esteem, and it's generally a healthy response. Remember, you already made your point by showing the error and suggesting ways to avoid the mistake in the future (as discussed in Chapter 11 on coaching). Leave it at that.

SELF-EFFICACY

As discussed in Chapter 13, self-efficacy is more situation specific than self-esteem, and so it fluctuates more readily. Job-specific feedback should actually affect only one's perception of what's needed to complete a particular task successfully. It should not influence feelings of general self-worth. Keep in mind, though, that repeated negative feedback can have a cumulative effect, chipping away at an individual's self-worth.

ACHIEVABLE TASKS. What makes for a "can do" attitude? Personal perception is the key. A supervisor, parent, or teacher might believe he or she has provided everything needed to complete a task successfully. However, the employee, child, or student might not think so. Hence, the importance of asking, "Do you have what you need?" We're checking for feelings of self-efficacy. This is easier said than done, because people often hesistate to admit they are incompetent. Really, who likes to say, "I can't do it?" Instead, we want to maintain the appearance of self-efficacy.

I have often found it necessary to ask open-ended questions of students to whom I give assignments, in order to assess whether they are prepared to get the job done. In large classes, however, such probing for feelings of self-efficacy is impossible. As a result, many students get left behind in the learning process (frequently because they skipped classes or an important reading assignment). As they get farther and farther behind in my class, their low self-efficacy is supported by the self-fulfilling prophecy and diminished optimism. Sometimes this leads to "give-up behavior" and feelings of helplessness. All too often, these students withdraw from my class or resign themselves to receiving a low grade.

Figure 14.2 reflects the need to focus on "small wins" when assigning tasks and communicating performance feedback. Of course, the kind of situation depicted in Figure 14.2 requires one-on-one observation and feedback in which the individual's initial competencies can

Figure 14.2. Starting small reduces errors and builds confidence.

be assessed. Then successively more difficult performance steps can be designed for the learner. The key is to reduce the probability that the learner will make an error and feel lowered effectiveness or self-efficacy. Celebrating small-win accomplishments builds self-efficacy and enables support from the self-fulfilling prophecy.

FOCUS ON THE POSITIVE. Many of the strategies I've presented for improving behaviors and person-states include a basic principle—focus on the positive. Whether attempting to build our own self-efficacy or that of others, success needs to be emphasized over failure. Thus, whenever we have the opportunity to teach others or give them feedback, we must look for small-win accomplishments and give genuine approval before commenting on ways to improve. This approach is easier said than done.

Failures are easier to spot than successes. They stick out and disrupt the flow. That's why most teachers give rather consistent negative attention to students who disrupt the classroom, while giving only limited positive attention to students who remain on task and go with the flow. Furthermore, many of us have been conditioned (unknowingly) to believe negative consequences (penalties) work better than positive consequences (rewards) to influence behavior change.

Figure 14.3 illustrates how natural variation in behavior can lead to a belief that penalties have more impact on behavior than rewards. If people receive rewards for their superior performance and penalties for their inferior performance, natural variation in performance (or regression to average performance) can give the impression that penalties are more powerful than rewards. In other words, the natural tendency to do less than superior work over time can mask the behavioral impact of a reward given for superior performance. On the other hand, the natural tendency to improve following inferior performance can give the impression that prior punishment improved the performance.

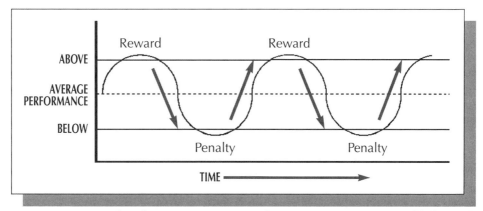

Figure 14.3. Natural performance variations make it easier to see a presumed behavioral impact of negative rather than positive consequences.

Figure 14.4. Look for the silver lining to build self-efficacy and self-esteem.

In summary, normal circumstances make it relatively difficult to focus on the positive. Mistakes are more noticeable than "go-with-the-flow" successes, and natural regression to average performance can develop a faulty belief that negative consequences have more behavioral impact than positive consequences. However, considering the impact consequences have on internal person states (especially self-efficacy and self-esteem), a positive consequence (like praise and social approval) is always preferable to a negative consequence (like criticism or ridicule). Indeed, as illustrated in Figure 14.4, it's often critical to search for admirable aspects of a situation and emphasize these in verbal interaction.

PERSONAL CONTROL

Most people find a healthy imbalance between internal (personal) and external control. They essentially believe, "I'm responsible for the good things that happen to me, but bad luck or uncontrollable factors are responsible for the bad things." Thus, people are apt to attribute injuries to rotten luck and beyond their control, especially when they happen to them. (Recall our discussion of attributional bias in Chapter 6.) This is the prime reason industrial safety must focus on process achievements rather than failure (injury) outcomes. To achieve a Total Safety Culture, people must believe they have personal control over the safety of their organization.

Employees at my seminars on active caring have listed a number of ways to increase perceptions of personal control, including:

1. Setting short-term goals and tracking progress toward long-term accomplishment

2. Offering frequent rewarding and correcting feedback for process activities rather than only for outcomes

3. Providing opportunities to set personal goals, teach others, and chart "small wins"

4. Teaching employees basic behavior-change intervention strategies (especially feedback and recognition procedures)

5. Providing people time and resources to develop, implement and evaluate intervention programs

6. Showing employees how to graph daily records of baseline, intervention, and follow-up data

7. Posting response feedback graphs of group performance

Figure 14.5 illustrates humorously a personal control perspective. Obviously, this is an extreme and unrealistic scenario. But wouldn't it be nice if people would attempt to take personal control of safety issues at their industrial sites with the same passion and commitment some individuals have for their golf game? I believe differences in perceived personal control for safety versus golf are largely due to contrasting scoring procedures.

Suppose you couldn't receive direct and immediate feedback about your golf game. That is, each time you hit a golf ball you wore a blindfold and couldn't see where the ball landed. Even when putting on the greens, you are blindfolded and can't tell whether your ball goes into the cup. Imagine also that you don't receive a score per hole or per game. However, you do receive negative feedback whenever your ball lands in a sandtrap. Under these circumstances, would you feel "in control" of your golf game? Would you attribute balls hit into sandtraps to personal control or just bad luck? Would you continue playing golf, or give it up for an activity in which you can experience greater personal control?

Of course, the golf scenario I've asked you to imagine is far-fetched. But, isn't this the way it is for safety at many industrial sites? The primary evaluation tool used to rank companies and determine performance appraisals and bonuses is an outcome number (such as total recordable injury rate) which is quite remote from the daily plant processes people have control over. Without a scoring system that focuses on controllable processes (as discussed earlier in this text), safety will be viewed as beyond personal control. An injury is just "bad luck," analogous to hitting a golf ball in a sandtrap while blindfolded.

Obviously, we cannot have complete control over all factors contributing to an injury. That's

Figure 14.5. Sometimes we try extra hard to exert personal control.

why I think it's wrong to say "all injuries are preventable." However, there is much we can do within our own domain of influence, and we can prepare for factors outside our personal control. Thus, we take an umbrella to the golf course in case it rains, and we wear personal protective equipment in case we are exposed to risks beyond our domain of personal control. Likewise we protect our children for events beyond their control, as illustrated in Figure 14.6.

Figure 14.6. We can't control everything!

OPTIMISM

As discussed in Chapter 13, optimism results from thinking positively, avoiding negative thoughts, and expecting the best to happen. Anything that increases our self-efficacy should increase optimism. Also, if our personal control is strengthened we perceive more influence over our consequences. This gives us more reason to expect the best. Again, we see how the person states of self-efficacy, personal control and optimism are clearly intertwined. A change in one will likely influence the other two. Recall from Chapter 13 that simple events like finding a dime in a coin return, receiving a cookie, listening to soothing music, and being on a winning football team are sufficient to boost optimism and willingness to actively care. It's not necessary for a person to perceive personal control over a pleasant consequence for that consequence to build optimism and possible actively caring behavior.

BELONGINGNESS

Here are some common proposals given by my seminar discussion groups for creating and sustaining an atmosphere of belongingness among employees:

- Decrease the frequency of top-down directives and "quick-fix" programs
- Increase team-building discussions, group goal-setting and feedback, as well as group celebrations for both process and outcome achievements
- Use self-managed or self-directed work teams

When groups are given control over important matters like developing a safety observation and feedback process or a behavior-based incentive program, feelings of both empowerment and belongingness can be enhanced. When resources, opportunities, and talents enable team members to assert,

"We can make a difference," feelings of belongingness occur naturally. This leads to synergy, with the group achieving more than could be possible from members working independently (see Figure 14.7).

Social psychologists have studied conditions that influence the productivity of people working alone versus in groups. Results of this research are relevant to increasing actively caring behavior, because conditions that spur on group productivity and synergy also boost feelings of empowerment and belongingness. When group members experience synergy, they especially appreciate the group process and project pride in their output. This feeds a win/win perspective and more perceptions of group empowerment and belongingness. Obviously, a Total Safety Culture requires teamwork, belongingness, and synergy. Yet, as you'll see below, traditional approaches to safety management can throw a wet blanket on this process.

Figure 14.7. Higher goals can be reached through synergy.

SOCIAL LOAFING: The whole can be less than the sum of its parts. First, I need to digress for a moment to describe a phenomenon known as social loafing. In the 1930s, social psychologists measured the effort people exerted when pulling a rope in a simulated tug-of-war contest. Researchers measured the force exerted by each of eight subjects when pulling on the rope alone and when pulling on the rope as a team. If the group effort was greater than the sum of the individual efforts, synergy would have been shown. What do you think happened?

The eight subjects worked harder alone than as a team. In fact, the total pulling force of the group was only about half the total of the eight individual efforts. This finding was not a fluke. This phenomenon, termed *social loafing*, has been demonstrated in more than 50 experiments conducted in the United States, India, Thailand, Japan, and Taiwan. For example, blindfolded subjects were asked to shout or clap their hands as loud as they could while listening through headphones to the sound of loud shouting or hand clapping. Subjects told they were doing the clapping or shouting with others produced about one-third less noise than when they thought they were performing alone. So what contributed to the social loafing?

IDENTIFIABILITY: CAN I BE EVALUATED? When subjects were told the sound equipment could measure their individual clapping as well as the team effort, social loafing disappeared. Thus, when subjects believed their personal effort could be identified, they worked just as hard in the group as alone. When are individual efforts identifiable in the workplace? Are there more opportunities to recognize individual effort in production and quality aspects of the job than for safety?

One thing is certain. The typical approach to measuring safety achievement—by counting number of hours or days without an OSHA recordable or lost-time injury—does not provide much opportunity for individual recognition, unless a worker is injured. And in this case, the recognition is failure-based, which motivates behavior through fear rather than desire. So what happens? The injury is often covered up to avoid individual embarrassment or punishment. That's not the behavior we want to see.

RESPONSIBILITY: AM I NEEDED ON THIS JOB? Social loafing has been shown to increase when a person's efforts are duplicated by another person. In one experiment, for example, subjects worked in groups of four to report whenever a dot appeared in a particular quadrant of a video screen. Social loafing did not occur when subjects were assigned their own quadrant to watch. However, social loafing was found when all four subjects were asked to observe all four quadrants. Social loafing can be expected to increase as people's sense of personal responsibility decreases. When people believe their individual contribution is important, perhaps indispensable, for the team effort, synergy is more likely than social loafing.

Do your safety processes allow individuals to feel a sense of personal responsibility for safety achievements? A safety system that is evaluated only by numbers of injuries or workers' compensation costs does not provide individuals with feedback regarding their contributions. But a safety monitoring process that tallies the safe versus at-risk observations of team members can build a sense of individual responsibility for a group's safety record. Likewise, when teams develop interventions to reduce hazards and increase safe behaviors, personal responsibility for team progress is realized and social loafing reduced. Tracking the success of intervention efforts verifies perceptions of empowerment and promotes feelings of group belongingness.

INTERDEPENDENCE: WE NEED EACH OTHER. Research on social loafing suggests approaches for facilitating synergistic teamwork. Social loafing is reduced when group members know each other well and agree on common goals. Assessing personal performance against an objective standard or the performance of other team members also diminishes social loafing.

In safety management this can be achieved by tracking individual contributions to safety observations and intervention processes. For example, team members can conduct regular audits of safe versus at-risk conditions and behaviors, providing invaluable data for calculating group percentages of safe conditions and behaviors. Work teams can also develop intervention strategies to correct environmental hazards or motivate safe work practice, and then devise a process to monitor individual contributions to the process.

HOW LARGE IS YOUR GROUP? Most of us feel a sense of belongingness with our immediate family, and as discussed in Chapter 13, we usually don't hesitate to actively care for the safety and health of family members (see Figure 13.5). In the same vein, the more interdependence and belongingness we experience at work, the more likely we are to look out for coworkers' safety. But how large is your work group?

Obviously, a large "family" means more people come under the protective wing of actively caring behavior. So it makes sense to have an extended family at work. In a Total Safety Culture, everyone actively cares for everyone else's safety. In effect, everyone belongs to one group. Now what are the barriers to an extended work family?

Figure 14.8 depicts an obstacle, or win/lose perspective, that I come across all too often. The "we-they" attitude spun off by traditional management-labor differences often makes for a dysfunctional "family." It seems as if some unions attempt to justify their existence by focusing on disagreement, conflict, and mistrust between management and labor. For its part, management supports this "we-they" split with an alienating communications style that asserts its ultimate power and control. I've seen management memos, for example, that might have been well-intentioned, but were written in top-down, control language that sounded like an adult talking to a child.

Through actively caring, and enhancing self-esteem, empowerment, and belongingness, we can bring down the "us versus them" walls that entrap a work culture. Active caring spreads mutual trust and interdependence throughout the culture. In a Total Safety Culture, everyone benefits from each individual's efforts.

Does actively caring imply the elimination of unions? No. But it might suggest altered visions and mission statements for organized labor groups. Labor

Figure 14.8. Win/lose inhibits teamwork and synergy.

unions can certainly help enhance the five person states that facilitate actively caring behavior. To do this, they need to work with management from a win/win perspective that appreciates interdependence and the power of synergy.

At the same time, managers need to relinquish their hold on the "control buttons" of operations and processes that workers can manage themselves, perhaps through self-directed work teams. A truly "empowered work force" is one trusted by managers and supervisors to get the job done without direct supervision. Obviously, this can't happen overnight, but a solid foundation is cemented when the five actively caring personal states are strengthened.

Directly Increasing Actively Caring Behaviors

You can treat actively caring behavior just like any other target behavior. Many interventions that increase the occurrence of safe work behaviors can be used to boost the frequency of actively caring behaviors. The four chapters in Part Three covered principles and procedures for directly influencing behavior. You'll recall that the techniques were classified as activators and consequences, with activators considered directive or instructional, and consequences being motivational. Let's take up that discussion again, as it applies to actively caring behavior.

EDUCATION AND TRAINING

There is some evidence that educating people about the barriers inhibiting actively caring behavior will increase acts of caring. The impact of education should be even more dramatic if some of the concepts discussed in Chapter 12 are taught. It seems particularly useful to explain that actively caring behavior can be proactive or reactive, direct or indirect, and focused on the environment, internal person-states, and behavior. Also, it should be taught that direct, proactive, and behavior-focused active caring is most challenging—and most useful to prevent injuries.

CONSEQUENCES FOR ACTIVELY CARING

Rewards should increase actively caring behavior. This comes from the basic operant learning principle that behavior increases following positive consequences. But it's perhaps even more important not to respond negatively when observing a caring act. A negative reaction could make that person avoid a subsequent opportunity to actively care for safety. I made a similar point when discussing safety coaching in Chapter 11. A person's reaction to a safety coach could determine whether or not that person goes out of their way to coach again for safety.

The "Actively Caring Thank-You Card" described in Chapter 10 is a rather generic technique to boost caring in the workplace. You'll recall that employees used these cards to thank their colleagues for going beyond the call of duty for the safety of others. Some cards included a space for the intervention agent to define the act of caring. Some cards could be exchanged for inexpensive rewards, like a beverage in the company cafeteria. Other cards had peel-off stickers for public display. In some applications, each card was worth 25 cents or 50 cents toward corporate contributions to local charities.

The contributions to charity idea promotes active caring in more than one way. First, those more ready to actively care, perhaps because of high levels of the internal person states, will find these "charity thank-you cards" more rewarding. These people not only set admirable examples for others, they're also likely to become champions of the process and the actively caring concept. Also, public displays of the charitable contributions can help get others involved in the process.

The Reciprocity Principle

Some sociologists, anthropologists, and moral philosophers consider reciprocity a universal norm that motivates a good deal of interpersonal behavior. Simply put, people are expected to help those who have helped them. You can expect people to comply with your request if you have done a favor for them. This is the principle behind Covey's (1989) claim that we need to make deposits in another person's emotional bank account before making a withdrawal.

I witnessed the reciprocity norm rather dramatically when I worked in the Virginia prison system in the mid-1970s. Several inmates used gifts and other forms of personal assistance to influence other inmates. I knew an inmate, for example, who went to great lengths to place a gift in the cell of a new inmate he wanted to influence. Accepting such a gift meant that the target inmate, actually a victim, was now obliged to return a favor in order to "save face."

In a similar vein, my father-in-law had a favorite expression whenever someone offered to do him a favor. He would say, "I don't want to be obligated." Without knowing the principle, he was saying that he didn't want the favor because he would feel a duty to reciprocate.

Have you ever felt a little uncomfortable after someone did you a favor? I certainly have. I interpret my discomfort as the reciprocity principle in action. Another person's kind act makes me feel pressured to respond in kind. What does this mean for safety? I think it means we should look for opportunities to go out of our way for another person's safety. Doing this, we increase the likelihood they will help when we need them.

The workplace setting might enhance chances for reciprocity. Research shows that people are most likely to pay back individuals they expect to see again. Also, the more actively caring behavior one receives, the more such behavior they feel obligated to return. So when we actively care frequently for the safety of coworkers we see often, we'll obligate them to return in kind. This actually has far-reaching benefits.

GIFTS AREN'T FREE. Has someone snared your attention to hear a sales pitch after giving you a free gift? Have you ever felt obligated to contribute to a charity after receiving gummed individualized address labels and a stamped envelope for your check? Ever purchase food in a supermarket after eating a free sample? Do you feel obliged to buy something after using it for a 10-day "free" trial period?

If you answer "yes" to any of these questions, it's likely you've been influenced by the reciprocity principle. Many marketing or sales-promotion efforts count on this "free sample" gimmick to influence purchasing behavior. Does this justify distributing free safety gifts, such as pens, tee-shirts, caps, cups, and other trinkets? Yes, to some extent, but you have to take into account perceptions. How special is the gift? Was the gift given to a select group of people, or was it distributed to everyone? Does the gift or its delivery represent significant sacrifice in money, time, or effort? Can it be purchased elsewhere, or does its safety slogan make it special?

A "special" safety gift—as perceived by the recipient—will trigger more acts of caring in response. Remember, too, that the way a gift is bestowed can make all the difference in the world. The labels and slogans linked with it can influence the amount and kind of responsive action. If the gift is presented to represent the actively caring safety leadership expected from a "special" group of workers, a certain type of reciprocity is activated. People will tell themselves that they are considered leaders, and they need to justify this label by going beyond the call of duty for others. And if they have learned about the various categories of actively caring behavior, they know the best way to lead is by taking action that is direct, proactive, and focused on supporting safe behavior or correcting at-risk behavior.

DOOR-IN-THE-FACE: START BIG AND RETREAT. Suppose the plant safety director pulls you aside and asks you to chair the safety steering committee for the next two years. This request seems outrageous, given your other commitments and the fact you never even served on the committee. You say, "Thanks, but no thanks!" The safety director says he understands, and then asks if you'd be willing to serve on the committee. Research shows that because the safety director "backed down" from his first request, you'll feel subtle pressure to make a similar concession—to reciprocate—and agree to the second, less demanding assignment.

Commitment and Consistency

Robert Cialdini refers to commitment and consistency as an influence mechanism lying deep within us, directing many of our actions. It reflects our motivation to be, and appear to be, consistent. "Once we make a choice or take a stand, we will encounter personal and interpersonal pressures to behave consistently with that commitment" (Cialdini, 1993, p. 51). Professor Cialdini suggests that the pressure comes from three basic sources:

- Society values consistency within people
- Consistent conduct benefits daily existence
- A consistent orientation allows us to take shortcuts when processing information and making decisions: We don't have to stop to consider everything involved; instead, we fall back on our prior commitment or decision and act accordingly to remain true to ourselves

PUBLIC AND VOLUNTARY COMMITMENT. The "Safe Behavior Promise Card" described in Chapter 9 derives its power to influence from the commitment and consistency principle. When people sign their name to a promise card they commit to behaving in a certain way. And then they act in a way that is consistent with their commitment.

Commitments are most effective, or influential, when they are visible, require some amount of effort, and are perceived to be voluntary, not coerced. It makes sense, then, to have employees state a public rather than private commitment to actively care for safety, and to have them sign their name to a promise card rather than merely raise their hand. It's critically important for those making a pledge to believe they did it voluntarily. In reality, decisions to make a public commitment are dramatically influenced by external activators and consequences, including peer pressure. But if people sell themselves on the idea that they made a personal choice, consistency is likely to follow the commitment.

FOOT-IN-THE-DOOR: START SMALL AND BUILD. This strategy follows directly from the commitment and consistency principle. To be consistent, a person who follows a small request will likely comply with a larger request later. During the Korean War, the Chinese communists used this technique on American prisoners by gradually escalating their demands, which started with a few harmless requests. First, prisoners were persuaded to speak or write trivial statements. Then they were urged to copy or create statements that criticized American capitalism. Eventually, the prisoners participated in group discussions of the advantages of communism, wrote self-criticisms, and gave public confessions of their wrong-doing.

Research has found this "start small and build" strategy succeeds in boosting product sales, monetary contributions to charities, and blood donations. This foot-in-the-door technique only works when people go along with the first small request. If a person says "No" right away, he or she might find it even easier to resist subsequent, more important requests. So if your first call for actively caring behavior is shot down, you didn't start small enough. Be prepared to retreat to something less demanding and build reciprocity from there.

THROWING A CURVE BALL: RAISING THE STAKES LATER. This technique of "throwing a curve" occurs when you persuade someone to make a decision or commitment because of the relatively low stakes involved. Then you raise the level of involvement required. For example, being a safety steering committee member isn't asking too much if meetings are held only once a month. But after attending the first two safety meetings, the stakes are raised. More meetings are requested for a special project. Because of the commitment and consistency principle, the individual will likely stick with the original decision and remain an active member of the committee.

This compliance tactic is almost legendary among car dealers. A customer agrees to a special purchase price, say $800 below all other competitors. Then the price is raised. A number of reasons—we should say excuses—are given. The sales manager won't approve the deal. Certain options were not included in the special offer. The manager decreased the value of the customer's trade-in. But the ploy usually works. Customers agreeing to the special price usually don't flinch when thrown the curve because reneging might suggest a lack of consistency or failure to fulfill an obligation, even though the obligation is only imaginary. Often customers develop their own reasons—or excuses—to justify their initial choice and the additional costs.

SOME INFLUENCE TECHNIQUES CAN STIFLE TRUST

Throwing the curve ball raises a critical issue when trying to increase actively caring behavior. How do you feel when someone throws you a curve? Let's talk about trust for a moment.

Do you trust the waiter who brings you an expensive wine list—only after you've been seated and made selections from the food menu? It probably depends on whether you believe this sequence of events was done intentionally to get you to buy more. Similarly, you might not dislike or mistrust the car salesman who jacks up the price unless you suspect that his advertised price was just a lure to get you in the showroom. In other words, our trust, appreciation, or respect for people might fall off considerably if we believe they intentionally tricked or deceived us into modifying our attitude or behavior. Of course, there may be no harm done if the result is clearly for our own good, as for our health or safety, *and* we realize this.

Some of the influence strategies reviewed here are more likely to raise suspiciousness and reduce interpersonal trust. The curve ball and door-in-the-face techniques are usually most difficult to pull off. But reciprocity might do the most harm if the recipient believes the kindness or favor is a self-serving manipulation done *only* to force feelings of indebtedness, as is often the case in prison cultures.

Trying to increase actively caring behavior is not self-serving. We're trying to build a safety culture that benefits everyone. This must be made clear. Once the purpose of using the influence techniques described here is understood, interpersonal trust or mutual respect between agents and recipients of interventions will not diminish but may actually grow.

In Conclusion

This chapter is probably the most important in terms of practical application. Integrating information from the prior chapters, I've tried to address this critical challenge: Continuous safety improvement leading to a Total Safety Culture requires people *to actively care for others as well as themselves.* Research-derived procedures to increase the frequency of actively caring behavior throughout a culture have been discussed. Some of these influence techniques indirectly increase actively caring behavior by benefiting the person states that facilitate one's willingness to care. Other influence strategies target behaviors directly.

Indirect strategies are deduced from the actively caring model explained in Chapter 13. Any procedure that increases a person's self-esteem, perception of empowerment—including self-efficacy, personal control, and optimism—or sense of belongingness or group cohesion will indirectly benefit active caring. A number of communication techniques enhance more than one of these states simultaneously, particularly actively listening to others for feelings and giving genuine praise for accomplishments. There are barriers to focusing on the positive, and we discussed them in the hope that awareness will help overcome the obstacles.

We need only reflect on our own lives to appreciate the power of choice, and how the perception of choice and personal control makes us more motivated, involved, and committed. Choice activates and sustains actively caring behavior.

Perceptions of belongingness are important, too. They increase when groups are given control over important decisions and receive genuine recognition for accomplishments. Synergy is the ultimate outcome of belongingness and win/win group involvement. It occurs when group interdependence produces more than what's possible from going it alone. We reviewed barriers to belongingness and synergy, specifically, social loafing.

The behavior-change techniques detailed in Chapters 9 and 10—from setting SMART goals and signing promise cards to offering soon, certain, and positive consequences for the right behavior—can be used to enhance those actively caring person states. They can also directly increase actively caring behavior. Education and direct participation in actively caring projects foster the behavior we're seeking, as well.

The interpersonal influence principles of reciprocity and commitment/consistency as they apply to our everyday decisions and behaviors were discussed. The commitment and consistency principle is behind the success of safe behavior promise cards, the foot-in-the-door technique ("start small and build"), and throwing a curve ball ("raise the stakes later"). These principles and specific strategies can be applied to directly boost actively caring behaviors.

Finally, it's critical to realize that social influence strategies can reduce respect or trust between people if they are applied within a context of win/lose or top-down control. On the other hand, when the purpose of the influence technique is clearly to increase actively caring behavior and improve the safety and health of everyone involved, a win/win climate is evident and mutual respect and trust is nurtured.

Part Five

Putting It All Together

Chapter 15

Evaluation for Continuous Improvement

Continuous improvement demands proper evaluation. This chapter explains how to evaluate the impact of safety interventions. The principles described here will make you a smarter consumer of marketed safety programs, and help you evaluate a customized intervention process.

What gets measured gets done; what gets measured and rewarded gets done well.

Larry Hansen (1994) used these words in his *Professional Safety* article on managing occupational safety (p. 41). You've probably heard words to this effect, indeed they are key to any continuous improvement effort. But there's a problem with how workplace safety is traditionally measured. As I indicated earlier in Chapter 3, too much weight is given to outcome numbers that people cannot control directly. People must be held accountable for results they can control. Yet corporations, divisions, plants, and departments are often ranked according to abstract outcome numbers like the total recordable injury rate. And these rankings often determine bonus rewards or penalties.

What behavior improves when safety awards are based only on an injury rate? If employees can link their daily activities to safety results, then celebrating reduced injury rates can be useful, even motivating. But it's critical to recognize the behaviors, procedures, and processes that led to fewer injuries or lower workers' compensation costs.

If you don't focus on the real causes of improvement, you run the risk of actually demotivating the folks deserving recognition. Employees might think continuous improvement is caused by luck or chance—events beyond personal control. This can lead to feelings of apathy or helplessness, as I discussed earlier (see Chapters 12, 13, and 14). If we want employees to work for continuous improvement, we need to recognize and reward the "right stuff." And this requires the right kind of measurement procedures.

Measuring Performance

Deming (1991) admonished his audiences for ranking people, departments, and organizations. In fact, he recommended that grades and performance appraisals be abolished completely from education and business. In his words, "The fact is that performance appraisal, management by the numbers, M.B.O., and work standards have already devastated Western industry...the annual rating of performance has devastated Western industry...Western management has for too long focused on the end product (Deming, 1986, July, p.1).

Part of Dr. Deming's rationale comes from the fact that standard approaches to measuring academic and work performance are often subjective, relative, and not clearly related to individual behavior. Teachers and professors, for example, use contrived distribution curves and cutoffs to assure only designated percentages of students can attain certain grades. Knowledge tests are necessarily biased and imperfect assessment devices, and are often only remotely linked to specific behaviors within a student's control—including attending class, taking notes, reading the textbook, and studying material on a regular basis. I've known many demotivated students who felt their daily efforts were overshadowed by the emphasis on exams.

If "what gets measured gets done," what are we measuring when we measure performance? From a systems standpoint, performance means output. In the context of psychology, performance means behavior, or output from a human system. We can't discuss measurement and evaluation without an operational definition of performance. Let's examine different definitions more closely.

INDIVIDUAL PERFORMANCE. The first fact I learned in my first psychology course was that psychology is the study of the individual. Later, each psychology course I took in graduate school examined some aspect of individual performance (the dependent measure) as it was influenced by particular environmental or experimenter manipulations (the independent variables). For example, in my courses on learning I studied the effects of prior experiences and reinforcement history on individual performance. In social psychology, I learned how other people influence an individual's performance; and in physiological psychology, I learned how specific changes to one's nervous system through drugs, electrical stimulation, or surgery influenced individual performance. Today, in my management and organizational psychology courses I teach students how an individual's work performance can be affected by training techniques, the three-term contingency (activator-behavior-consequence), and differences in a person's knowledge, skills, ability, attitude, and personality.

SYSTEM PERFORMANCE. At a four-day Deming seminar in 1991, I learned that many corporate leaders and experts in the field of organizational management—including safety management—define performance from an industrial systems perspective. The second afternoon was devoted to the famous red/white bead game, a demonstration to convince the audience that individual performance is a relatively insignificant determinant of organizational, or system, performance. Dr. Deming asked for ten volunteers from the more than 600 attendees. He assigned six volunteers the job of "willing worker," two other volunteers became "inspectors" of each worker's output, another volunteer played the role of "quality control supervisor" or inspector of the inspectors' work, and the tenth volunteer was the "recorder."

After assigning work duties, Dr. Deming mixed 800 small red beads with 3,200 white beads—all the same size—in a box. He showed the "willing workers" how to scoop out beads from this box using a beveled paddle that had indentations for 50 beads. Each scoop of the paddle was considered a day's production. The inspectors counted the numbers of red and white beads on each worker's paddle, and after a reliability check by the bead inspector, the number of red beads (considered mistakes or defects) was recorded and displayed publicly on an overhead projector by the recorder.

Dr. Deming played the role of the corporate executive officer, and he urged the workers to produce no defects (red beads). When workers scooped fewer than 10 red beads, they were praised by the CEO. Scooping more than 15 red beads, however, resulted in severe criticism and an exhortation to do better, otherwise "we will go out of business." Dr. Deming reacted to each scoop of beads as a product of individual performance; if performance outcome was poor, he again demonstrated the correct bead-sampling procedure, or gave corrective feedback to a worker during the sampling process. It's noteworthy that regression to the mean resulted in good individual performance getting worse and poor individual performance getting better. As discussed in Chapter 14, this can give the impression that correcting feedback works better than rewarding feedback.

The sampling, inspecting, reliability checking, recording, public graphing of results, and corrective action for defects continued for ten samples (representing ten days) per worker. This portion of the demonstration lasted more than an hour, and led logically to Deming's concluding statements that "the actual number of red beads scooped by each worker was out of that worker's control. The workers only *delivered* defects. Management, which controls the system, caused the defects through system design."

This demonstration has profound lessons for achieving a Total Safety Culture. As I've discussed earlier, the typical evaluation procedure used by both the government and private sectors to judge the safety record of companies is based on organizational or system performance—"the total recordable injury rate" or number of "OSHA recordables"—uncontrollable by most individual workers. Individuals are not only held accountable, perhaps even disciplined, for their own injuries but also for the injuries of others, even employees outside their immediate work group. They are often blamed for injuries caused by a number of factors outside their control, such as at-risk environmental conditions or equipment, excessive workloads, system contingencies causing a fast work pace, and a culture that supports "rugged individualism" and thwarts group cohesion and a "brothers/sisters keeper" perspective.

In July, 1992, Dr. Deming wrote me this explanation to clarify his teaching intentions with his red/white bead demonstration.

> *Currently, management works under the assumption that people and not the systems they work in are responsible for performance. We therefore reward and punish people but the system they work in remains unchanged. The point is not that differences between individuals are unimportant in and of themselves (but) that focus on individual differences alone yields possibilities for improvement that are trivial compared with transformation of the entire system that they work in. In fact, only individuals can change a given system to improve performance. (W. E. Deming, personal communication, July 11, 1992)*

SYSTEM VS. INDIVIDUAL PERFORMANCE. In the Fall of 1991, a demonstration in my graduate course in behavior and systems management illustrated profoundly the role of individual performance in organizational output. It revealed the need to consider both individual and system-level factors when evaluating safety achievement. One of my students, Mario Beruvides, directed a probability "game" that, to his surprise and my delight, illustrated the significant and unexpected impact of individual performance factors. Playing cards were used instead of beads to illustrate the importance of system-level performance.

Mario asked four volunteers and myself to play the role of workers for a hypothetical company, "Geller Inc., Non-Face Card Manufacturers." He gave us each a sealed deck of cards, and instructed us to open them, remove the two jokers and instruction cards and shuffle them thoroughly. Then we placed our decks face down on the seminar table, and per Mario's instructions, drew five cards from the top of our decks to represent a day's work. Face cards were counted as defects or errors, and the number of face cards per "willing worker" was recorded and displayed to the class on a "production log."

As in Deming's bead game, the CEO, played by Mario, urged the workers to care for the welfare of the company and produce as few defects as possible. Defects from each worker were recorded, summarized, and displayed by a student assigned as "supervisor." The five-card drawing and recording continued for five trials, representing a work week. After each trial, the CEO reprimanded the workers whose rate of defects exceeded the average and praised those who had just one or zero defects. To the delight of the class and surprise of the CEO, my drawing always included the most face cards, or defects, and I always received the most criticism.

Before the sixth round of drawings, marking the second work week, the CEO unveiled a company-wide incentive plan: Whenever the face cards from a group of five "hands" are combined within 15 seconds to form a pair or three or more of a kind, as in a poker hand, the face cards will not be counted as defects. On each trial, I continued to draw the most face cards, usually three or more, and I immediately took charge of organizing the face cards from the other "hands" into poker combinations. This leadership and teamwork decreased the number of defects per individual dramatically. Now, my defect rate was at least as low as any other worker's.

During the lively classroom discussion that ensued, Mario admitted that our results were remarkably different than those obtained by the management training and development specialist who invented the card-game analog of Deming's bead demonstration (Storey, 1989). In a completely random system, one "worker" is not supposed to always produce the most defects. Plus, Chris Storey reported that the incentive condition should not reduce defects, even over five draws, presumably because some "workers" are reluctant to trade cards due to lack of experience, knowledge, or leadership.

While discussing reasons for the unexpected results from "Geller, Inc." I made a provocative discovery—I had been drawing cards from a pinochle deck containing 50 percent face cards. Mario had inadvertently purchased one pinochle deck, and happened to give me that deck by chance. This serendipitous innovation in Mario Beruvides' demonstration resulted in profound awareness of the power in individual differences.

Mario admitted his preset expectation that the game would show random, uncontrollable variation. His anticipation to find only "common cause" prevented him from recognizing "special cause" contributions from individuals. And he was caught completely off guard by the success of the incentive program. The students identified individual performance factors that made the incentive program effective, particularly the special effects of the pinochle deck and its chance assignment to the assertive class instructor. This led us to discuss how companies can benefit from appropriate employee selection and placement, education and training, interpersonal communication, and individual and group recognition processes. We also saw the need to shift from our usual preoccupation with individual performance and attend to group and system-wide performance as well.

A Total Safety Culture, then, requires us to balance how we measure and manage both organizational performance and individual behavior. The challenge is to accurately attribute change in safety performance. Is it due to individual behavior, groups, or the system? Individuals should only be held directly responsible for their own safety performance; teams should be held accountable for outcomes directly related to their team performance.

Continuous improvement in organizational safety performance, using the yardstick of OSHA recordables, requires improvements in the system as a whole, of which individual employees play an integral part—both as individuals and members of work teams. As I've discussed earlier, holding individuals responsible for safety performance outside their perceived control develops attitudes and perceptions, such as apathy, helplessness, and pessimism, that interfere with both personal and organizational safety improvement.

Developing a Comprehensive Evaluation Process

In April and August of 1995, I had the pleasure of working with a panel of evaluation experts to develop a set of measurement guidelines for the National Safety Council and the Centers for Disease Control and Prevention. Our mission was to develop a handbook of practical guidelines that field personnel can use to evaluate the impact of an intervention to improve safety on family and commercial farms (see Steel, 1996).

We agreed that evaluation was essential to hold people accountable for achieving program objectives. This particularly pertains to those developing and implementing the intervention. The evaluation process should measure whether intervention procedures: a) are consistent with relevant principles and mission statements of the organization; b) reach the desired audience and are implemented as planned; and c) are efficient and effective. In sum, a complete evaluation process should assess how an intervention was conceived, designed, and implemented, and how efficient and effective it was.

Figure 15.1 summarizes our panel's deliberations on how to evaluate intervention impact. It integrates the primary issues discussed so far in this chapter. Lower levels of the hierarchy represent process activities needed to improve the higher-level outcomes of a safer environment and ultimate injury reduction. Immediate causes of injury reduction are changes in environment or behavior—or both.

Let's pause for a moment to consider the cause-and-effect connection between process and outcome. Behavior can be viewed as an outcome, shaped by a process that pursues changes in employee knowledge, perceptions, or attitudes. In this case, the behavior-change goal depends on employee participation in an intervention aimed at influencing a person state. Of course, intervention processes can be designed to circumvent person states and directly change behavior to reduce injuries, as discussed in Part Three of this text.

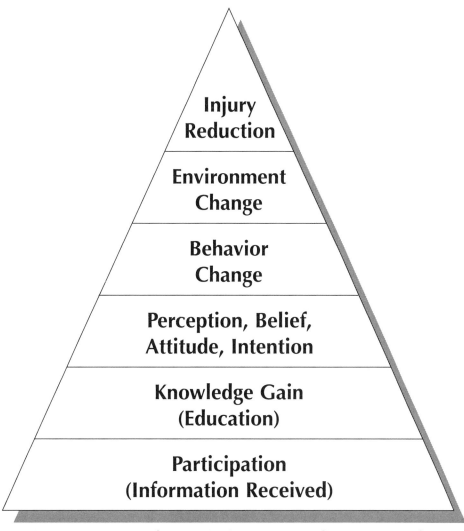

Figure 15.1. Measures of intervention impact vary according to remoteness from immediate injury causation.

Figure 15.1 points out the relativity of process and outcome. An education process, for example, can lead to behavior change (an outcome), while a process to change behaviors might result in outcome changes to the environment, say improved housekeeping. And completing a work process in a safer environment can affect the ultimate outcome—fewer injuries.

Figure 15.1 also reflects the three basic areas requiring attention for injury prevention: environment, behavior, and person. This is, of course, the "Safety Triad" introduced in Chapter 2 to categorize intervention strategies, and referred to later in Chapter 12 to classify different types of actively caring behaviors. The hierarchical levels of intervention impact in Figure 15.1 reflect one or more of these three domains and suggest a particular approach to measurement. Changes in environments and behaviors can be assessed directly through systematic observation, but changes in knowledge, perceptions, beliefs, attitudes, and intentions are only accessible indirectly through survey techniques, usually questionnaires. Let's review these three basic areas of evaluation.

WHAT TO MEASURE?

Most safety interventions focus on either environmental conditions—including engineering controls—or human conditions, as reflected in employees' perceptions, attitudes, or behaviors. It might seem reasonable to us to evaluate change only in the area we've targeted—environment, behavior, or person state. When the target is corporate "culture," employee perceptions or attitudes are typically evaluated. If behavior change is the focus, then behaviors are observed and analyzed in terms of their frequency, rate, duration, or percentage of occurrence, as reviewed in Chapter 8. When environments or engineering technologies are evaluated, mechanical, electrical, chemical, or structural measurements are taken.

I've heard culture-change consultants advocate perception surveys in place of environmental audits. And presentations on behavior-based safety emphasize direct observations of work practices, often in lieu of the subjective evaluation of personal perceptions and attitudes. But given the need for employees to "feel good" about a behavior-based safety process, and their need to participate continuously in managing and monitoring this process, I think it's obvious that we need to check perceptions and attitudes and ongoing behaviors. For example, a comprehensive evaluation of a simple change in equipment design should probably include an assessment of relevant human factors, like employees' work behaviors around the new equipment and their attitudes and perceptions regarding the equipment change.

I hope you can see that a comprehensive evaluation for safety requires a three-way audit process covering environmental conditions, safety-related behaviors, and person states such as perceptions, attitudes, beliefs, and intentions.

EVALUATING ENVIRONMENTAL CONDITIONS. In many ways, environmental audits are the easiest and most acceptable type of evaluation. In fact, regular environmental or housekeeping audits are already standard practice at most companies. But these evaluations can often be improved by involving more employees in designing audit forms, conducting systematic and regular assessments, and posting the results in relevant work areas.

Figure 15.2 depicts a generic environmental checklist for safety that can be used to graph for public display the percentage of safe conditions and the percentage of potential corrective actions taken for at-risk tools, equipment, or operating conditions. My associates at Safety Performance Solutions typically teach work teams the rationale behind the environmental checklist, and

Observer:_____ Date:_____ Time:_____
Department:_____ Building:_____ Floor:_____ Area:_____

OPERATING CONDITIONS/TOOLS & EQUIPMENT	SAFE	AT-RISK	CORRECTIVE ACTIONS TAKEN*
Electrical wiring (properly enclosed)			
Air nozzles (limited to 30 psi)			
Chemicals (exposure concern)			
Eyewash station			
Emergency shower			
Barricades (in place where necessary)			
Storage of materials (neat/safe)			
Hazard communication labels (appropriate)			
Floors (dry)			
Exits, aisles, sidewalks and walkways (clear of debris)			
Lighting (adequate)			
Housekeeping (satisfactory)			
Tools (safe operating condition)			
Guards (adequate and in place)			
Fire extinguisher (in appropriate location)			
Forklift (safe operation)			
Forklift driver (license in possession)			
Tow truck (safe operation)			
Tow truck driver (license in possession)			
Chairs (safe condition)			
Totals			

Percent Safe Conditions: $\dfrac{\text{Total Safe Observations}}{\text{Total Safe Observations + At-Risk Observations}}$ x 100 = _____%

Percent Corrective Actions Taken: $\dfrac{\text{Total Corrective Actions Taken}}{\text{Total At-Risk Observations}}$ x 100 = _____%

*Please list and define the corrective actions taken on back of this sheet.

Figure 15.2. An environmental checklist to evaluate the safety of tools, equipment, and operating conditions.

then assist them in applying the checklist in their plant. Employees then customize the checklist and graphing procedures for their particular work areas. Regular audits and feedback sessions increase accountability for environmental factors that can be changed to prevent an injury.

EVALUATING WORK PRACTICES. The systematic auditing of work practices was the theme of Chapters 8 and 11. In Chapter 8, I introduced the overall *DO IT* process: *D*efine target behaviors, *O*bserve target behaviors, *I*ntervene to increase safe behavior or decrease at-risk behavior, and *T*est (or evaluate) the impact of your intervention. Also in Chapter 8, we discussed how to develop two types of observation checklists—a generic version for basic work practices applicable anywhere, such as prescribed lifting techniques and use of certain personal protective equipment; and a job-specific checklist for particular tasks, like the safe driving checklist my daughter and I developed.

The coaching process detailed in Chapter 11 also discussed how to develop and apply both generic and job-specific checklists for one-on-one observation and feedback sessions with coworkers. As mentioned above, using a behavioral checklist to observe and evaluate ongoing work practices is the type of performance appraisal that can lead to continuous improvement.

EVALUATING PERSON FACTORS

As discussed earlier in this text, person factors refer to subjective or internal aspects of people. They are reflected in commonly used terms like attitude, perception, feeling, intention, value, intelligence, cognitive style, and personality differences. You can find many surveys that measure specific person factors of target populations ranging from children to adults. Some of these factors are presumed to be *traits,* others are considered *states*. It's important to understand the difference when you consider the evaluation potential of a particular survey.

PERSON TRAITS. Theoretically, traits are relatively permanent characteristics of people; they don't vary much over time or across situations. Since traits are relatively permanent, questionnaires that measure them cannot gauge the impact or progress of a culture-change intervention. Trait measures serve as a tool to teach individual differences, but in safety management their application is limited to selecting people for certain job assignments. This is very risky, though, and not very effective, as discussed in Chapter 1.

PERSON STATES. Person states are characteristics that can change from moment to moment, depending on situations and personal interactions (as discussed in Chapter 13). When our goals are thwarted, for example, we can be in a state of frustration. When experiences lead us to believe we have little control over events around us, we can be in a state of apathy or helplessness. Person states can influence behaviors. Frustration, for example, often provokes aggressive behavior; and perceptions of helplessness inhibit constructive behavior or facilitate inactivity.

In contrast, certain life experiences can affect positive person states, such as optimism, personal control, self-confidence, and belongingness. These, in

HIGHLY AGREE

AGREE

NOT SURE

DISAGREE

HIGHLY DISAGREE

	1	2	3	4	5
1. The risk level of my job concerns me quite a bit.	1	2	3	4	5
2. When told about safety hazards, supervisors are appreciative and try to correct them quickly.	1	2	3	4	5
3. My immediate supervisor is well informed about relevant safety issues.	1	2	3	4	5
4. It is the responsibility of each employee to seek out opportunities to prevent injury.	1	2	3	4	5
5. At my plant, work productivity and quality usually have a higher priority than work safety.	1	2	3	4	5
6. The managers in my plant really care about safety and try to reduce risk levels as much as possible.	1	2	3	4	5
7. When I see a potential safety hazard (e.g., oil spill), I am willing to correct it myself if possible.	1	2	3	4	5
8. Management places most of the blame for an accident on the injured employee.	1	2	3	4	5
9. "Near misses" are consistently reported and investigated at our plant.	1	2	3	4	5
10. I am willing to warn my coworkers about working unsafely.	1	2	3	4	5
11. Employees seen behaving unsafely in my department are usually given corrective feedback by their coworkers.	1	2	3	4	5
12. Compared to other plants, I think mine is rather risky.	1	2	3	4	5
13. Working safely is the Number One priority in my plant.	1	2	3	4	5
14. I have received adequate job safety training.	1	2	3	4	5
15. Many first-aid cases in my plant go unreported.	1	2	3	4	5
16. Information needed to work safely is made available to all employees.	1	2	3	4	5
17. Management here seems genuinely interested in reducing injury rates.	1	2	3	4	5
18. Safety audits are conducted regularly in my department to check the use of personal protective equipment.	1	2	3	4	5
19. I know how to do my job safely.	1	2	3	4	5
20. Most employees in my group would not feel comfortable if their work practices were observed and recorded by a coworker.	1	2	3	4	5

Figure 15.3. Questionnaire items designed to measure personal perceptions of organizational safety. (Selected from the Safety Culture Survey developed by Safety Performance Solutions, Inc.)

turn, boost constructive behavior. This was the indirect approach to increasing actively caring behavior discussed in Chapter 14.

Measures of person states can be used to evaluate perceptions of culture change, and to pinpoint areas of a culture that need special intervention attention. Like most culture surveys, our Safety Culture Survey asks participants to answer questions on a five-point continuum (from highly disagree to highly agree) about their perceptions of the safety culture. Issues include the perceived amount of management support for safety, the willingness of employees to correct at-risk situations and look out for the safety of coworkers, the perceived risk level of the participant's job, and the nature of interpersonal consequences following an injury.

Our survey also measures factors that increase one's willingness to actively care for another person's safety. These include self-esteem, belongingness, and empowerment, as detailed in Chapter 13. Sample items from our survey that measure the actively caring person states are given in Chapter 13. They were adapted from professional measures of these characteristics, and have been evaluated for reliability and validity. Figure 15.3 contains twenty items from the safety perception and attitude portion of our survey. You'll note nothing very special about the items in this scale. They ask employees to react to straightforward statements about safety management and improvement.

You could compare employees' reactions to the items in Figure 15.3 before and after implementing a safety improvement process. Studying reactions prior to an intervention helps identify issues or work areas needing special attention. This information can lead you to choose a particular intervention approach, or customize one. Data from a baseline perception survey might even indicate that a culture is not ready for a given intervention process, suggesting the need for more education and discussion to get employees to "buy in."

Evaluating Costs and Benefits

The benefits of an occupational safety program are illustrated dramatically by the cost-benefit evaluation of a company-wide safety-belt program initiated at Ford Motor Company. This corporate-wide program was initiated in the spring of 1984, before any state passed a mandatory seat-belt use law. Safety committees at the Ford plants developed their own safety-belt programs, based on the behavior-change principles described in Part 3 of this text. The combination of activator interventions like awareness sessions, incentives, and commitment pledge cards with rewarding consequences more than tripled safety-belt use among Ford employees. After only one year, this increase saved the lives of at least eight employees, spared about 400 others from serious injury, and reduced corporate costs by ten million dollars. After the second year of increased belt use, the benefits more than doubled, amounting to direct monetary savings of approximately $22 million.

This cost-benefit analysis was invaluable in sustaining top-management support for the campaign at Ford. It also provided motivating feedback to the many employees who developed, managed or evaluated the plant-based interventions. It's important to maintain records of direct and indirect costs associated with injuries—and with injury prevention—even if these calculations are only estimates. Comparing estimates with the costs of implementing and maintaining particular safety programs illustrates specific benefits and justifies continued program support, especially if you can show that the program has substantially reduced injury frequency and costs.

When calculating program costs, you should document every expense, including: promotional materials, teaching aids, evaluation supplies, rewards, media expenditures, and wages paid for program assistance. Employees' time used to plan, present, evaluate, or participate in the program should be estimated, even if the time is voluntary—for example, on evenings or weekends. If you are comprehensive when calculating program costs, then you are justified in estimating the numerous direct and indirect costs resulting from a job-related injury.

Injury records should be consulted before and after an intervention process is implemented in order to show savings from fewer work-related injuries. Direct costs that should be calculated per injury include:

- Wages paid to absent employees (workers' compensation)
- Property damage
- Medical expenses
- Physical and vocational rehabilitation costs
- Survivor benefits

These direct costs may be the proverbial tip of the iceberg when considering the indirect or hidden costs of business disruptions caused by a lost-time injury. These costs might be difficult or impossible to calculate, but you should try to estimate costs in these categories:

- Overtime pay used to cover the work of an injured employee
- Scheduling work tasks to cover for an injured employee
- Additional administrative hassles, extra wages, training time, and inefficient work associated with temporary replacements
- Special costs of losing a skilled employee
- Extra time from work supervisors to schedule shift changes, temporary replacements, or employee training necessitated by the absence of an injured worker

- Retraining and readjustment for employees returning to work after an extended absence (a permanent disability from the injury may call for a new job assignment)
- Special costs for extensive recruitment procedures and on-the-job-training for permanent replacements for injured employees who don't return to work or who return with a permanent disability
- Special administrative costs to investigate and document the incident and medical treatments for compliance with state workers' compensation laws and with other state and federal regulations, such as OSHA standards

Considering the direct and indirect costs to a company from work-related injury is overwhelming in at least two respects. It's certainly an intimidating chore to estimate these costs, and it's stunning to think that the corporate losses from one employee's injury can be so great. These dollars clearly justify considerable intervention activity. Just anticipating the negative consequences from a work-related injury should motivate support and participation in proactive efforts. This is the first critical step in "selling" safety—the theme of the next chapter.

You Can't Measure Everything

Deming (1991, 1992) condemned grades and performance appraisals because they provide a limited picture of an individual's contributions and potential. And they might constrain the number and type of interventions used to improve the quality of a work culture. If, for example, the only procedures implemented to improve safety are those that allow for objective measurement, the number and quality of safety interventions is severely restricted. In Chapter 14, for example, I discussed a number of ways to increase actively caring behaviors directly, through applications of learning and social influence principles, and indirectly, through improving the five personal states that increase willingness to actively care. It's impractical and impossible to measure the impact of many of these interventions. Should we avoid doing so just because we can't measure their occurrence and impact?

Dr. Deming explained there are many things we should do for continuous improvement without attempting to measure their impact. We shouldn't do these things only to influence performance indicators, but because they are the right things to do for people. You might never be able to measure the impact of treating an employee with special respect and dignity, but you do it anyway. Such treatment may, in fact, contribute to achieving a Total Safety Culture but you'll never know it. Likewise, you'll never know how many injuries you prevent with proactively caring behaviors, and you'll never know how much actively caring behavior you'll promote by taking even small steps to increase coworkers' self-esteem, empowerment, and belongingness. You need to continue doing these things anyway. Many things that can't be measured and rewarded still need to get done.

In Conclusion

At the start of this text, I explained the fallacy of basing decisions on common sense. Rather than adopt intervention programs that sound good, we need to use procedures that work. But how do we know what works? Of course, you know the answer to this question. Only through rigorous program evaluation can we know whether an intervention is worth pursuing. Now comes the more difficult question. What kind of program evaluation is most appropriate for a particular situation?

Actually, every chapter of this text has addressed program evaluation in one way or another. Early on I explained the need for achievement-oriented methods to keep score of your safety efforts. This enables people to consider safety in the same work-to-achieve context as production and quality. This implies, of course, the need for program evaluation numbers people can understand and learn from. This is how evaluation leads to continuous improvement.

Throughout this text, I have referred to published research in order to justify psychological principles or recommendations for intervention procedures. Information presented is founded on rigorous evaluation, not common sense. Evaluation techniques used in published research are indeed more rigorous and complex in terms of reliability, validity, and statistical analyses than those needed for continuous improvement of real-world safety programs. But the basic principles and issues presented in this chapter are relevant to both researchers (seeking to contribute to professional scholarship) and practitioners (seeking continuous improvement of an intervention process).

To publish their findings, researchers need to demonstrate reliability and validity of their measures and find statistical significance. However, they can and do ignore several evaluation principles presented in this chapter. For example, their measures typically: a) target only one dimension (environment, behavior, or person factors); b) are short-term (applied for a limited number of experimental phases); c) are subjected to statistical transformations and analyses that take substantial time to complete and are not readily understood by the average person; and d) often do not include a cost-benefit analysis. You see, reports of their procedures and results only need to be understood and appreciated by a select, often esoteric, group of professionals who specialize in the particular issue or problem addressed by the research.

But you can't overlook most of the basic principles when evaluating practical interventions to achieve continuous improvement. Data collection procedures and statistical analyses often can be less rigorous. But safety practitioners need to address several important issues often bypassed by professional researchers. Specifically, they need to: a) define the level of performance targeted by the intervention, while appreciating limitations in attacking individual versus organizational performance; b) use measures for the three dimensions of safety improvement—environment, behavior, and person factors; c) apply process measures periodically over the long term, especially checks on environmental conditions and work practices; d) include a cost-benefit analysis to justify continued intervention and evaluation efforts; and e) keep score with numbers that are both meaningful to all program participants and provide direction to refine interventions. These last two principles are critical to meet the challenge addressed in the next chapter—obtaining and maintaining support for an effective intervention process.

Chapter 16

Obtaining and Maintaining Involvement

You can't effectively put to use the principles in this book without sustained support from both managers and employees. This chapter focuses on ways to initiate and maintain that support, including ways to promote leadership, build commitment and involvement, expand the scope of interventions, and reduce active resistance.

Culture change is never quick, never easy. The "quick-fix" illustrated in Figure 16.1 is clearly ridiculous. As absurd as this notion is, it comes to us naturally. We want speedy solutions to difficult challenges. And it's easy to lose patience, enthusiasm, and optimism along the way. After all, our society demands immediate gratification—just look at all the movies and television shows that begin with dramatic problems and come to happy endings within 30 to 90 minutes. Plus, the faster we solve any problem, the sooner we experience rewarding consequences.

Figure 16.1. There's no quick fix solution to culture change.

Of course, we know the insinuation depicted in Figure 16.1 is preposterous. Common sense tells us, correctly this time, that fairy-tale solutions are utopian and unrealistic. It makes much more sense to plan and work for the small wins that come from SMART goals (Chapter 10). We need to remember the well-known answer to the silly question, "How do you eat an elephant?" One bite at a time!

This chapter brings us to the point where we start pulling things together. I want to discuss the broad challenge of initiating a culture-change process aimed at achieving a Total Safety Culture. First come general guidelines for starting a process and maintaining support. Then, I address concepts of leadership, communication, and resistance. You won't find step-by-step cookbook procedures here; a generic recipe is just not available. Instead, take the principles and procedures presented in prior chapters, add the information found here, and you'll be well on your way to an innovative experience in safety improvement.

Starting the Process

MANAGEMENT SUPPORT

You can't do without it. How many times have you heard: "Whatever management really pushes and supports will happen." And, implicitly, "Whatever upper management does not push and does not support will fail." If managers emphasize housekeeping, quality control or cost-reduction, improvements in these areas are likely to follow. Strong top-down support, involvement, and commitment alone won't make a campaign succeed, but they are essential ingredients. Plus, management and labor must collaborate to make the process work.

CREATING A SAFETY STEERING COMMITTEE

A Safety Steering Committee (SSC) plays a critical role in developing a Total Safety Culture, providing policy-making, oversight, and general support. All this is simply more than any one person can handle. At the start, there are at least two functions that the commitee serves:

Content function: groups typically produce better ideas, and more creative problem solutions than individuals working alone

Process function: group-based solutions or decisions lead to more commitment and enthusiastic involvement than individual solutions or decisions handed down

Before creating a Safety Steering Committee, it's important to look at the existing committee structure in your organization. You don't want to duplicate the efforts of the safety department, or some other relevant standing committee. For example, a current employee committee might be able to take on the responsibility of coordinating efforts to achieve a Total Safety Culture.

Careful planning is needed to determine:

- What is the mission of the Safety Steering Committee?
- What is the vision for SSC operation?
- What are the group's limitations or restrictions?
- What are the priorities?
- Who should be on it?

DEVELOPING EVALUATION PROCEDURES

"Did it do any good?" This is the central question that the Safety Steering Committee must be prepared to answer about any intervention. Chapter 15 offered guidelines for evaluating impact. At the start of an evaluation process, these questions need to be answered:

- What indicators should we look at—behaviors, attitudes, opinions, or outcomes?
- When should we measure?
- What types of data should we collect and analyze?
- What is the cost of this evaluation process?
- How should we summarize and display results?

SETTING UP AN EDUCATION AND TRAINING PROCESS[2]

Ensuring that employees learn key principles and procedures to improve safety is a major responsibility for the committee. At minimum, the following elements should be incorporated into planning an effective education and training program:[3]

- Develop education content and procedures
- Plan the education and training process
- Plan for follow-up sessions
- Identify and prepare instructors
- Measure the impact of the program

---

2. I distinguish between educating and training, with *educating* referring to teaching the principles or theory behind a process or set of procedures and *training* referring to teaching techniques or procedures with hands-on experience and behavioral feedback.

3. A process is a long-term continuous effort, mission, or set of procedures which might include specific programs with a beginning and end. Thus, a training process might include a particular training program that is refined or updated on a regular basis.

Let's discuss each of these elements in a bit more detail, keeping in mind that my suggestions need to be customized at the plant level to get the most "bang for your buck."

IDENTIFYING AND PREPARING INSTRUCTORS. Selecting the "right" instructors is critical because teaching is at the heart of effective education and training. If the teachers do a poor job, they undermine other training tools such as videos and booklets. You should consider these factors when choosing instructors:

- Prior experience in educating or training
- Current level of teaching ability (aptitude and achievement)
- Credibility with employees to be educated
- Level of motivation and interest in doing the instruction
- Prior familiarity with psychology, especially behavior-based principles
- Belief that the principles and procedures can help achieve a Total Safety Culture

To help selected teachers prepare for their task, they will need to: a) understand the principles and relevant procedures in this text so they can represent them accurately to the participants; b) feel comfortable with the specific process of the plant-wide education and training which, ideally, they will help to develop; c) practice basic communication skills; d) demonstrate leadership at meetings; and e) learn how to facilitate discussions—particularly if they have little experience at stand-up training and small-group meetings.

In-house training staff can help volunteers build their teaching skills. These topics are often covered in various corporate training programs. Plus, instructional programs are readily available from outside vendors or consultants.[4]

DEVELOPING TRAINING CONTENT. When considering the specific topics to include in each instructional session, you need to address these points:

- What are the specific goals of a particular session—for example, to teach principles, train procedures, or build commitment and motivation?
- In addition to this text, what sources are available for relevant content and support, such as local case studies of behavior-based safety?
- What relevant films, videotapes, and other instructional materials are available?[5]

---

4. My associates at Safety Performance Solutions (SPS) provide training and guide organizations through the entire process reviewed here. For more information write Safety Performance Solutions, 1007 N. Main Street, Blacksburg, VA 24060 or call (540) 951-SAFE (7233).

5. A videotape training series (including four videotapes, workbooks, and facilitator guides) has recently been developed from the information presented in this text. The series is called "Actively Caring for Safety," with the four separate programs entitled: 1) Motivating Safe Behavior, 2) Implementing Behavior-Based Safety, 3) Coaching Safe Behavior, and 4) Making Safety Incentives Work. For more information write Tel-A-Train, Inc., PO Box 4752, Chattanooga, TN 37405 or call 1-800-251-6018.

- To meet session goals, what key content points should be covered, and in what order?
- What specific plant facts, statistics, and case studies can be incorporated into the session?
- How much information can be covered effectively in one session?

Involve your selected instructors as much as possible in developing the specific education and training plan.

PLANNING THE INSTRUCTIONAL PROCESS. "Everything was covered but nobody paid much attention." This is a common complaint about education or training. Translation: Good content is important, but not sufficient. You've got to "package" your content and present it in a way that hits home with participants. Obviously, you want them to practice what is preached. Most instructors know their material; the challenge lies in conveying that knowledge. How do you get your message across? Here are some points to consider:

- An interactive/participative approach is typically more effective than a "top down" lecture coming from the podium
- Like a good pitcher, change speeds. Don't rely on one pitch, one way of presenting information
- It is easier to involve small groups of participants than large ones
- Regardless of the main objective of a session, some initial awareness raising makes participants more receptive to the content
- Integrate demonstrations into the program
- If possible, have participants practice the skill taught, with appropriate feedback in the classroom or on the job
- Resolve those administrative questions, including where and when the instructional sessions should be held, and how long they should take

EVALUATING EFFECTIVENESS. As covered in Chapter 15, you can't overlook measurement. At this point in the overall intervention process, you need to determine impact of education and training on participants' knowledge, skill, and attitude.

Knowledge of content can be assessed the old-fashioned way, through written tests given at the end of a session. Skills can be evaluated by systematic behavioral observation in the classroom or out in the workplace. Participants' reactions to the session can be measured with brief questionnaires. For this you should keep in mind:

- Brevity
- Choice and complexity of wording
- Combining objective rating and written comments

SUSTAINING THE PROCESS

Continued upper-management support. Many safety programs get a big send-off, only to drop off the radar screen. Then it's "out of sight, out of mind" as some new program is pushed. This is why safety is often derided for its "flavor of the month" approach. So how do we sustain a safety process? (Note that I prefer the concept of "process" to "program" here because processes flow on while programs begin and end.) First and foremost we need continued and visible support from top management. If management endorses the process on an ongoing basis, it can become integrated into normal plant operations.

The Safety Steering Commitee needs to work hard to convince managers that their commitment is fundamental to the process—not only to get it going, but to keep it going. Here are some thoughts on maintaining that all-important backing of top managers:

> First, you need to gain access to upper management. Identify a manager to champion your cause in the executive offices. Bring him or her into the loop, ask them to attend all committee meetings.
>
> Keep managers informed. Submit committee reports to them on a regular basis. But don't overwhelm busy managers with minutiae.
>
> Ask for input, keep managers involved. Solicit comments and "concerns" about the process you have underway. Of course you should be doing this with all levels of the organization to create a top-to-bottom sense of ownership.
>
> Promote and market your efforts. Publish articles and announcements about the safety process in the employee newsletter on a regular basis.
>
> Keep at it. Identify the benefits of your process and continue to "sell" them to upper management.

AWARENESS SUPPORTS: ACTIVATORS

Reminders of various kinds can be extremely helpful to keep managers, supervisors, and employees aware and involved in your safety process. Vary these support devices and by all means keep them coming. In addition to management commitment and activity, other sustaining process drivers are:

- Environmental supports, such as signs, slogans, newsletters, articles
- Incentive supports, especially publicity about reward strategies
- Promise cards to pledge behavioral commitment
- Social support at group meetings and social gatherings

PERFORMANCE FEEDBACK: CONSEQUENCES

As you know, a cornerstone of the DO IT approach to motivating behavior change is the principle that behavior is influenced by its consequences.

But in most plants there are no clear-cut consequences for performing safe acts. Sometimes there are actually subtle rewards for *failing* to follow proper safety procedures. Employees might save time and avoid discomfort by not wearing appropriate protection, for instance. As detailed in Chapter 11, *rewarding* feedback should be given following safe behaviors, and *correcting* feedback should be given when at-risk behaviors are observed.

Effective performance feedback is, of course, closely linked to measurement and evaluation—the "test" phase of DO IT. As with measurement, performance feedback should be given systematically and consistently. Ideally, feedback regarding safe behavior should occur at several levels, including plant-wide (for overall organizational performance), across shifts or work groups (for group performance), and individually (for personal performance).

Finally, it's important to acknowledge supervisors and others for supporting a safety achievement effort. As discussed in Chapter 11, safety coaches must feel appreciated for their actively caring efforts. They need to be recognized for their competence at giving supportive and corrective feedback to others.

Without clear behavioral consequences, performance tends to drift, usually in the direction of minimal effort. If you want to sustain the energy it takes to be a safety coach, you need to recognize their work. You have to convince top managers that coaches need their support, as well.

ONGOING MEASUREMENT AND EVALUATION

You can't maintain and improve a safety process unless you regularly measure its impact. As discussed in Chapters 10 and 11, even the mere act of tracking a given set of behaviors usually improves the performance of those behaviors. Performance tracking should be done continuously, and questions to be resolved include:

- Who conducts ongoing measurement?
- What will be measured?
- How frequently should the program's impact be measured?
- What types of evaluation data should be taken (for example, frequency counts or percentages)?
- How will data be summarized and tabulated?
- What are the costs of the evaluation?
- What are the benefits—real and potential?

FOLLOW-UP INSTRUCTION/BOOSTER SESSIONS

Even with ongoing support, a comprehensive safety achievement process cannot succeed without carefully planned follow-up instruction. From time to time education and training content must be updated to reflect changes in plant conditions, the use of new machines or protective equipment, and the like. Don't delay in keeping pace with change; what is being taught should correspond exactly to current plant conditions. Follow-up education and training includes these issues:

- When and how should basic instruction be repeated for new employees?
- What are the objectives of follow-up instruction?
- How should new material be integrated?
- How often should follow-up sessions occur?
- What should be the content?
- How should the material be presented—videotapes, lecture, on-the-job training, discussion groups?
- How can monthly safety talks be used as boosters or activators?

TROUBLESHOOTING AND FINE-TUNING

Once a safety achievement process is up and running, the Safety Steering Commitee confronts the responsibility of fine-tuning the procedures. This is based on ongoing evaluations. If the process is going to be sustained, employees and managers must perceive it as current—state of the art—in terms of content, and also adaptable and responsive. It simply cannot be "frozen," nor left unattended. You can't "wind up" a process at the start and expect it to run forever like that battery-powered bunny. Some keys to fine-tuning include:

Discuss the impact: Is it working? What do the data from participants' reaction sheets, as well as other measures, tell us?

Identify strengths and weaknesses: Based on the data, which elements should be kept, changed (and how), or replaced?

Cope with change: Be sure all those affected by the process are fully informed about changes when they occur. Ideally, all participants should be actively involved in troubleshooting and fine-tuning interventions.

Cultivating Continuous Support

Starting a safety achievement process and maintaining it for long-term improvement requires the three essential support processes depicted in Figure 16.2. Leaders are needed to champion new principles and procedures. In fact, leadership makes the difference between a "flavor of the month" safety initiative and a long-term continuous improvement process.

WHERE ARE THE SAFETY LEADERS?

First, we have to find the leaders. Who are they? The traditional definition of one person exerting influence over a group doesn't quite work for safety. Ask

1. Leaders communicate effectively.
2. Leaders recognize desired performance.
3. Recognition is communicated effectively.
4. Leaders recognize desired performance effectively through a variety of communication channels.

Figure 16.2. Continuous improvement depends on three support processes.

any safety manager who has been expected to do it all. To achieve a Total Safety Culture, everyone needs to accept a leadership role in reducing injuries. Everyone feels responsible for safety and goes beyond the call of duty to protect others. This requires leadership skills: giving supportive feedback for another person's safe behavior and constructive feedback for at-risk behavior.

Psychologists have studied leadership rigorously for over half a century in an attempt to define the traits and styles of good leaders. Still, many questions remain unanswered, making leadership more an art than a science. But several decades of research has turned up some important answers, which we'll now apply to safety.

PASSION. The most successful leaders show energy, desire, passion, enthusiasm, and constant ambition to achieve. Passion to achieve a Total Safety Culture can be fueled by clarifying goals and tracking progress. Put a positive spin on safety, make it something to be achieved—not losses to be controlled. Then employees will be motivated to achieve shared safety goals just like they work toward production and quality goals. Marking progress leads to the genuine belief that the process works. This fires up employees to continue the process.

HONESTY AND INTEGRITY. Effective leaders are open and trustworthy. A Total Safety Culture depends on open interpersonal feedback. This obviously requires honesty, integrity, and trust. It's quite useful for work groups to discuss ways to nurture these qualities in their culture. Take a look at certain environmental conditions, policies, and behaviors. Some arouse suspicions of hidden agendas, politics, and selfish aims. You can work to eliminate some of these trust-busters by first identifying them, discussing their purpose, and devising alternatives.

MOTIVATION. Since most people really care about reducing personal injuries, even to people they don't know, the motivation to lead others will spread naturally throughout a work culture when people believe they can have personal control over injuries. This occurs when they learn effective techniques to prevent injuries (as presented in Part Three of this text) and feel empowered to apply them (as covered in Part Four).

SELF-CONFIDENCE. Effective leaders trust in their own abilities to achieve. Education helps convince people they can achieve, but they need ongoing support and recognition for their efforts. For example, the self-confidence needed to give safety feedback can be initiated with appropriate education and training, and can be maintained with coaching, communication, and recognition. Notice that enhancing the three empowerment factors—self-efficacy, personal control, and optimism—as discussed in Chapter 14, builds self-confidence.

THINKING SKILLS. Successful leaders can integrate large amounts of information, interpret it objectively and coherently, and act decisively as a result. Constructive thinking skills evolve among team members when objective data are collected on the progress of safety interventions and used to refine or expand these processes and develop new ones.

When teams work through the DO IT process (as presented in Part Three), participants develop skills to evaluate behavioral data and use the information to make intervention decisions. This is basic scientific thinking, the key to substituting profound knowledge for common sense. This critical thinking leads to special expertise.

EXPERTISE. To achieve a Total Safety Culture, everyone needs to understand the principles behind policies, rules, and interventions to improve safety. When employees teach these principles to coworkers, they develop the level of profound knowledge and expertise needed for exemplary leadership.

FLEXIBILITY. Successful leaders size up a situation, and adjust their style accordingly. At times some groups and circumstances call for firm direction—an autocratic style. At other times, the same people might work better under a nondirective, hands-off approach—a democratic style. The best leaders are good at assessing people and situations, and then matching their behavior to fit the need.

IN SUMMARY. So here we have seven personal qualities that distinguish leaders from followers. It's common to consider these characteristics permanent inborn traits. But a more optimistic perspective, one helpful to achieving a Total Safety Culture, is that employees can nurture these ideals in themselves and others. There is no quick and easy way to to do this. But the principles reviewed here should help you discriminate between action plans that facilitate versus inhibit safety leadership. And people already possessing leadership qualities must help bring them out in others.

COMMUNICATE TO SELL THE PROCESS

How we talk about safety influences whether people will contribute their leadership skills to a safety process. Indeed our language can determine acceptance or rejection of the process.

> *Words are magical in the way they affect the minds of those who use them...words have power to mold men's thinking, to canalize their feelings, to direct their willing and acting.*

This quote from Aldous Huxley's "Words and Their Meanings" (Hayakawa, 1978, p.2), reflects the power of words to shape our feelings, expectancies, attitudes and behaviors. When people use expressions like, "Say that enough times and you'll start to believe it," "Can't I talk you into it," or "Do as I say, not as I do," they acknowledge the influence of words on behavior.

I think some words we use in the safety and health field are counterproductive. Let me point out a few that should be eliminated from our everyday language if we want to "sell" safety and increase involvement to prevent injury.

"ACCIDENT" IMPLIES "CHANCE." The first definition of "accident" in my *New Merriam-Webster Dictionary* (1989) is "an event occurring by chance or unintentionally" (p. 23). Workplace accidents are usually unintentional, of course, but are they truly chance occurrences? There are usually specific controllable factors, such as changes in the environment, behaviors, and/or attitudes, that can prevent "accidents."

We want to develop the belief and expectation in our work culture that injuries can be prevented by controlling certain factors. "Accident," then, is the wrong word to use when referring to unintentional injuries. It reduces the number of people who believe with true conviction that personal injuries can be prevented.

RESTRAINTS DON'T INVITE USE. For almost two decades I've been urging transportation and safety professionals to stop using the terms "occupant restraints" and "child restraints" for vehicle safety belts and child safety seats.

These terms imply discomfort and lack of personal control, and fail to convey the devices' true function. "Seat belt" is better than "occupant restraint," but this popular term is not really adequate because it doesn't describe the function or appearance of today's lap-and-shoulder belts. We need to get into the habit of saying "safety belt" and "child safety device."

PRIORITY OR VALUE. Priority implies importance and a sense of urgency, and safety professionals are often quick to say "safety should be a priority." This seems appropriate, since my *New Merriam-Webster Dictionary* defines "priority" as "taking precedence logically or in importance" (p. 577). But everyday experience teaches us that priorities come and go. Depending upon the demands of the moment, one priority often gets shifted for another. Do we really want to put safety on such shifting ground?

I believe a Total Safety Culture requires safety to be pursued as a value. The relevant definition of "value" in my *New Merriam-Webster Dictionary* is "something (as a principle or ideal) intrinsically valuable or desirable" (p. 800). Safety should be a "value" that employees bring to every job, regardless of the ongoing priorities or task requirements.

DON'T SAY "BEHAVIOR MODIFICATION." Over recent years I've seen "behavior modification" used many times for titles of safety presentations at regional and national conferences. I've heard trainers, consultants, and employees use the term to describe behavior-based safety. In fact, I've often been introduced at conferences as a specialist in "behavior modification." This is the wrong choice of words to use if we want acceptance and involvement from the folks who are to be "modified." Who wants to be "modified"?

Actually, the term "behavior" alone carries negative associations for many—as in "Let's talk about your behavior last night"—but I can't see any way around using it. We need to teach and demonstrate the benefits of focusing on behaviors, especially desirable behaviors. But we don't have to link "modification" with behavior; it only adds to the negative feelings.

I hope the basic message is clear. We need to understand that our language can activate feelings and even behaviors we don't want. Figure 16.3 illustrates what I mean. If we want to communicate in order to "sell" the process, we need to consider how our language will be perceived by the "customer."

Figure 16.3. The words we use can increase participation or resistance.

As we end this discussion on language, I'd like you to ponder the following word-attitude associations. What terms better contribute to employee involvement?

"Peer pressure" or "peer support"?

"Loss control managers" or "safety facilitators"?

"Compliance" or "accomplishment"?

"Meeting OSHA standards" or "fulfilling a corporate mission"?

"30 days without an injury" or "30 safe days"?

"I must meet this deadline" or "I choose to achieve another milestone"?

"I've got to do this" or "I get to do this"?

It's a good personal or group exercise to consider the ramifications of using these terms and phrases, and adding alternatives to this list is even more beneficial. But understanding the critical relationship between words, attitudes, and deeds is only half the battle. We need to change verbal habits, and this is easier said than done. When we communicate with greater passion and optimism about safety we'll attract more people to our safety mission. We'll also reduce resistance to change.

Overcoming Resistance to Change

"How do we deal with people who resist change?" "How do we get more people to participate?" I frequently hear these questions at training seminars and workshops. First, let's face reality: Change is unpleasant for many people, and some are apt to react poorly. Change often threatens our "comfort zones"—those predictable daily routines we like to control. In fact, it takes a certain amount of personal security and leadership to try something new. A certain kind of risk taking is needed to lead change, and some people want no part of exploring the unknown.

We've all been in unfamiliar situations where we're not sure how to act. We feel awkward and uncomfortable. If someone gives us direction, helps increase our sense of control, it's easier to adjust. We might even help others deal with the change. But without leaders and adequate tools to cope with change, we might retreat, withdraw from the situation, or even actively resist the change.

So how do we deal with resistance? Simply put, we should teach people the skills and give them the tools to handle change, plus support those who set the right examples. This seems logical and intuitive, but it doesn't always happen. Instead, managers too often try to identify the malcontents and discipline them for not participating.

Let's try to better understand resistance by considering one of the classic awkward situations thrust upon many of us—our first school dance. Remember it? For me, it was a high school homecoming dance in 1957.

Attending your first dance is like a rite of passage. If you were anything like me, you were a bit nervous about this change in your social world. You might have been prepared for it. Family, friends and teachers probably told you what to expect. Maybe you even had dance lessons. But these "tools" didn't make it any easier for some of us to participate. Not me, anyway. I didn't participate, but I wanted to. Before the dance, I practiced how to ask a girl to dance. I took four, two-hour dance lessons at an "Arthur Murray Dance Studio." I felt ready, but never once did I dance the entire night. I didn't feel too embarassed, though, because there were so many others not dancing. As was the custom, boys stood on one side of the gym and the girls on the other.

As illustrated in Figure 16.4, some kids were dancing and seemed to be having a great time. They danced almost every number, and tried to lure others out on the floor. I couldn't be enticed, though. I hung back in my comfort zone. But at least I was in the dance hall.

As illustrated in Figure 16.5, some students stood around in the parking lot, talking, drinking, and smoking cigarettes. These were the resisters. Some were active resisters. They stayed in their cars, never intending to enter the dance hall. Now and then, these guys started up their cars and cruised around town for awhile, and then returned to the parking lot. They'd persuade others to hop in their car, try some beer, smoke a cigarette, fool around, or cruise.

LEVELS OF PARTICIPATION. There are essentially five ways of reacting to change—call them levels of participation—and they were all on display at the dance. First, there are the true leaders who get totally involved. They are the innovators—those who view change as necessary, and an opportunity to improve. At the dance, they were the teenagers on the floor for almost every number. They had the most fun. They didn't necessarily know what they were

Figure 16.4. Different reactions to change can be seen at the high school dance.

Figure 16.5. Some actively resist change and others follow.

doing when it came to dancing, but they got out there and tried. They took a risk. They got totally involved and benefited most from the occasion. A dance might start with only a few of these "risk takers," but they often persuaded a number of others to get involved as the night wore on.

Some people want to change, but need direction and support. They're motivated to participate, but need models or leaders. At the dance, these were the kids who hung back at first. With a little encouragement they danced a few numbers. By the end of the evening you couldn't get them off the dance floor. They were now totally involved.

Most of us are at the third level of participation. We're ready to get involved, but we'll stay in our comfort zones until we're directed and motivated to participate. It might look like we're resisting change, but not really. Call us neutral when it comes to our attitude about change. We just aren't sure what to do. We need self-confidence that we can handle the change. And we need genuine support (positive recognition) when we try to participate. Once in a while, just getting started or "breaking the ice," is enough to turn a passive observer into an active participant. But for the most part, we stand on the sidelines and watch. This level of participation was represented by the boys and girls who lined each side of the gym.

TYPES OF RESISTANCE. The final two levels of participation are passive and active resistance. Passive resisters perceive change as a problem. They complain a lot. They're critical and untrusting of something new imposed on them. They seem to see only the negative side of a new program, policy, or challenge. They rationalize their position by gathering with others at their same level of nonparticipation, and they grumble and whine about proposed changes or about others who are participating in a change effort. But their whining and complaining usually stops when participation in a new process is clearly enjoyed by the majority. Passive resisters are followers, and they'll do what they see most people doing.

These are the teenagers who came to the dance because everyone else would be there, but they felt so insecure or anxious they didn't enter the building. They looked for others hanging around outside, and made fun of the silly dancing going on inside. Sometimes these nonparticipants ran into an active resister.

Fortunately, active resisters are few in number. But it doesn't take many of these characters to slow down a change process. These individuals view change as a threat or an opportunity to resist. They see any change effort that was not their idea as a potential loss of personal control, and they often exert countercontrol to assert their control or freedom. Most parents observe this phenomenon when their children reach the teens. Teenagers want to feel independent, and at times will disobey their parents' directions—break the rules—to gain a sense of independence or self-control.

We all feel overly controlled at times, and perhaps react to regain independence or assert personal freedom. Sometimes our reactions are not thoughtful, caring, or safe. But, active resisters feel the need to resist change, the status quo, or authority much of the time. This is partly because their contrary behavior brings them special attention—recognition for resisting.

Who gets the attention? Active resisters stick out and attract attention. Nonparticipants use them to rationalize their own commitment to comfort zones. Managers monitoring the workplace often hit them with discipline. But this can backfire. This makes the top-down control more obvious for resisters. Discipline builds their resentment of the system and makes it even less likely they will join the change process. For some individuals, disciplinary attention only fuels their burning desire to exert independence and resist change. As a result, they might become more vigorous in recruiting others to oppose change. As I discussed in Chapter 10, top-down discipline should be used sparingly if the ultimate purpose is total participation in an improvement process.

How were resistant teenagers brought inside to the dance? The harsh warnings of the school principal shouting from the steps didn't work; neither did the one-on-one confrontation between one of the adult chaperones and the "leader of the pack." Whenever I saw a resister come inside, it was always the result of urging by another teenager. Peer pressure (or peer support) is still the most powerful motivator of human behavior. Therefore, the best way to deal with resistance is usually to arrange for situations that enable or facilitate peer influence. Eventually, some of the resistant teenagers came into the building so they wouldn't miss something. As depicted in Figure 16.6, they saw from their remote comfort zones that the people inside were really enjoying themselves, and they chose to participate.

Figure 16.6. When a critical mass of the culture changes, others follow.

The dance party will become more enticing as more and more teenagers dance. To increase such active involvement, the right kind of encouragement and support is needed. Will motivational lectures from a teacher, counselor, concerned parent, or outside consultant make that happen? It might make a temporary difference, but not over the long haul. The best way to deal with nonparticipation is usually to set up situations that allow for peer influence. This could mean managers do nothing more than support the change process, and let peer pressure or support occur naturally.

POWER OF PEER INFLUENCE. So how do you facilitate peer influence? The best way is through empowerment, but this is easier said than done. You can give people more responsibility, such as the challenge to lead others in a change process, but they must *feel* responsible. As I discussed in Chapter 12, they need to have sufficient self-efficacy ("I can do it"), personal control ("I am in control"), and optimism ("I expect the best").

Some people already have sufficient self-efficacy, personal control, and optimism when assigned leadership responsibilities. As discussed in Chapter 12, they feel empowered ("I can make a difference"). Still, these individuals may need some basic training in communication, social recognition, and behavior-change principles. Others may lack one or more of the three person states that facilitate feelings of empowerment. So, in addition to education and training, they need a support system to build their sense of self-efficacy, personal control, and optimism.

As discussed in Chapter 14, there's no quick fix for increasing perceptions of empowerment. But if you ask people to define policies, settings, interactions, and contingencies that influence these three person states, you're on your way to developing action strategies to improve them. This in turn increases people's readiness to be empowered.

Remember the bottom line: Resistance to change is overcome most effectively through peer influence. This requires, of course, that people are willing to accept the leadership position of *change agent* (Chapter 11). Certain individuals naturally rise to the occasion and welcome opportunities to lead. Others may be committed to continuous improvement through change, but need some direction and encouragement from the natural leaders. Both of these groups need training and practice in behavior-change principles and social influence strategies. Then management needs to give these people direction and the opportunity to work with those folks standing on the sidelines in need of a nudge to leave their comfort zones.

It's usually best to ignore the resisters. Don't give them too many opportunities to say "no." The more often they publicly refuse, the more difficult it will be for them to change their minds and participate. So don't pressure these folks. Invite them to contribute whenever they feel ready to achieve success with those willing to try something new. When the majority buys in and eventually celebrates its accomplishments, resisters will choose to come on board. The key is for them to perceive that they chose to get involved rather than being forced by a top-down mandate.

My teenage daughter's recent experience at her first high-school dance allows me to use this analogy once more to make a final point. Karly was well prepared for the dance, but perhaps not for her first date. She was quite talented at performing the latest dances, she was wearing a new outfit, and she had planned to meet a number of her girlfriends who were also attending their first homecoming dance with a date.

After the football game, Karly was to meet her date at the dance. She arrived at the gym before him, and instead of waiting outside, she went on inside. She then got totally involved in the dance, heading out to the floor for just about every song with whomever was available and willing, boy or girl. In this situation, Karly was an innovator and a leader. She felt empowered enough to lure others onto the dance floor so they could join in the fun. When Karly's date finally arrived, he rushed up to her, led her off the dance floor, and admonished her for not waiting outside. Why should she have fun without him?

It's important to realize that even leaders need support for their leadership efforts. Those who benefit from a leader's inspiration or coaching should share their appreciation. Karly's date actually punished her for her initiative and total involvement.

Fortunately, Karly had sufficient self-efficacy, personal control, and available support from friends to ignore her date's reprimand. In fact, she went back to the dance floor and participated with the support group that had evolved before her date arrived. And she chose to make her first date with that boy her last. That empowerment was sure appreciated by her dad!

In Conclusion

This chapter began with a list of guidelines to initiate and sustain a culture-change process aimed at achieving a Total Safety Culture.

Three support processes were identified to maintain employees' long-term commitment and involvement in a culture-change effort—leadership, communication, and recognition. Each of these processes were discussed in various forms throughout previous chapters. Aspects covered in this chapter were: a) characteristics of effective leaders; b) safety language that increases resistance and should be avoided; and c) levels of resistance that can be influenced by leadership, communication, and recognition.

Psychologists find the best leaders are enthusiastic, honest, motivated, confident, analytical, informed, and flexible. Although it's common to see these characteristics described as permanent personality traits, it's certainly reasonable to assume that they can be increased through education, communication, recognition, and involvement in a behavior-based safety process. Thus, while it is useful to look for "natural" leaders when selecting members of a Safety Steering Committee, it's important to realize that leadership qualities could be suppressed in some people by their lack of empowerment or belongingness. New safety processes and eventual culture change might bring out leaders you didn't know existed in the workforce.

Involvement is key to so many aspects of building a Total Safety Culture, and it can be increased many ways. Start by getting into the habit of using more positive safety language. And focus less attention on the active resisters. There are five levels of involvement in any change effort that you should recognize: a) total involvement from innovators who see change as an opportunity to improve; b) individuals committed but not totally involved until direction and support are given; c) people, usually the majority, ready but on the sidelines until prodded and encouraged by others; d) doubters who see change as a problem and use learned helplessness and criticism as excuses to remain detached; and e) the active resisters who see change as an opportunity to resist, complain, and promote mistrust.

Active and passive resisters (categories d and e) should be ignored, if possible. Recognize and support those willing to try the new process. Employees totally involved in the process (category a) need to help individuals committed but not yet totally immersed (category b). Then these two groups can work with the majority (category c) who need examples to follow. You can see why it's important to cultivate leadership, communication, and recognition skills among the "true believers" in innovation. Turn these leaders loose and they'll be your best recruiters to build the base of support for a Total Safety Culture.

Chapter 17

Reviewing Fifty Key Principles of a Total Safety Culture

Fifty principles for understanding the human dynamics of safety summarize what this book is all about. When you use these principles to design, execute, evaluate and continuously improve interventions to change safety-related behaviors and attitudes, you're well on your way to achieving a Total Safety Culture.

"How should we translate these concepts into real-world application?"

"Would you please put your theory into procedures or practices we could follow in our plant?"

I heard questions like these at each of the Deming workshops I attended. They seemed to disappoint Dr. Deming (1991, 1992), who would assert that the purpose of the seminar was to teach theory and principles, not specific procedures. It was up to the participants to return to their own organizations and devise specific methods and procedures that fit their culture. Dr. Deming stressed the need to start with theory and then customize practices.

This text also downplays a "one-size-fits-all" solution. Packaged programs are not the answer to safety problems, though they can be found everywhere. Sure, the quick fix might work for a while. Short-term success follows a familiar pattern: Injuries reach unacceptable levels, management hires a consultant, employees react with interest, injuries go down, statistics improve. Improvement is not difficult if injury rates are bad to begin with. But as I discussed earlier, this superficial approach only improves the worst organizations for a limited period. For companies with a good safety record, it doesn't work even for the short term.

As Dr. Deming well knew, and as I've discussed throughout this text, lasting improvement is built on specific procedures that fit the culture of an organization. Outside consultants can be invaluable, teaching appropriate principles and facilitating the implementation process. But if most of the employees do not understand and believe in the principles to begin with, well-intentioned efforts never take root.

That's why throughout this book I've presented theory and principles from psychological research to help you design safety-improvement interventions. The successful applications of these principles that I've described in brief case studies are not intended as step-by-step procedures to follow, but rather examples to consider when customizing a process for your culture.

It all starts with theory. And in this final chapter I want to pull together 50 important principles that summarize the psychology of safety, and lay the groundwork for building a Total Safety Culture. I know 50 sounds like a long list, but don't worry, I'll be brief. And the principles will be familiar to you, having come from information already covered.

I hope you'll find the list useful as a review, and a starting point for developing your safety-enhancement process. Some of the principles focus on design and implementation. Others explain why we often fail in safety. Most can be used as guidelines for checking potential long-term benefits of a specific safety-improvement procedure. All help you appreciate the complex human dynamics of safety and health promotion.

This is not a priority list. Don't read anything into the order of principles. What I hope is that you will teach them to others. You *can* make a difference, and bring about constructive culture change.

PRINCIPLE 1: Safety should be internally—not externally—driven.

It's common to hear employees talk about safety in terms of OSHA—the Occupational Safety and Health Administration. It often seems they "do" safety more to satisfy the mandates of this outside regulatory agency than for themselves. This translates into perceptions of top-down control, and performing to avoid penalties rather than to achieve success.

Ownership, commitment, and proactive behaviors are more likely when we work toward our own goals, not the government's. As discussed in Chapter 16, how we define programs and activities can influence attitudes that shape involvement. It makes sense to talk about corporate safety as a mission owned and achieved by the very people it benefits.

PRINCIPLE 2: Culture change requires people to understand the principles and how to use them.

In Chapter 16 (footnote 3), I distinguished between education and training, and emphasized that long-term culture change requires both. Education focuses on theory or principles. Training gets into the specifics of how to turn principles into practical procedures. Role playing or one-to-one interaction is very important because participants get direct feedback on how they're executing procedures or processes.

PRINCIPLE 3: Champions of a Total Safety Culture will emanate from those who teach the principles and procedures.

When people teach, they "walk the talk" and become champions of change. After more than 20 years of safety consulting, it's clear to me that success depends on the presence of these leaders. I've seen no better way to develop champions of a campaign than to first teach relevant theory and method, then show how others can be instructed, and finally allow opportunities for colleagues and coworkers to teach each other.

PRINCIPLE 4: Leadership can be developed by teaching and demonstrating the characteristics of effective leaders.

Just because you believe in something doesn't guarantee you'll be an effective champion of the cause. Leaders have certain characteristics, as discussed in Chapter 16, which can be learned and cultivated in others. People need to understand the principles behind good leadership, and the behaviors that reflect good leadership qualities. You can also learn by observing the leadership skills of others. And when you see leaders in action, reward their exemplary behavior with feedback.

PRINCIPLE 5: Focus recognition, education, and training on people reluctant but willing, rather than on those resisting.

As discussed in Chapter 16, people resist change for many reasons. Some feel insecure leaving their comfort zones. Some mistrust any change in policy or practice that wasn't their idea. Others balk for the special attention they get by resisting. It's usually a waste of time trying to force change on these folks. In fact, resistance hardens as more pressure is applied.

It saves time to prioritize. First focus on those who "want to dance." The ones willing to get involved. Then turn these leaders loose on the folks who "came to the dance" but are reluctant to get involved. At least these people are willing to consider a change proposal. Peer instruction can cultivate change champions (Principle 3), as well as increase participation. When a critical mass of individuals gets involved and achieves success as a result of change, many initial resisters will join in—out of choice, not coercion.

PRINCIPLE 6: Giving people opportunities for choice can increase commitment, ownership, and involvement.

A basic reason for preferring the use of positive over negative consequences to motivate behavior (Chapter 10) is that people feel more free, they perceive more choice, when working to achieve rewards. As illustrated in numerous laboratory experiments and field applications, increasing perception of choice leads to more motivation and involvement in the process (Chapter 14).

Personal choice also implies personal control—enhancing empowerment and willingness to actively care for others (Chapter 14). It's important to realize that eliminating the perception of choice—by imposing a top-down mandate that restricts or constrains work behavior, for example—can sap feelings of ownership, commitment and empowerment, and inhibit involvement.

PRINCIPLE 7: A Total Safety Culture requires continuous attention to factors in three domains: environment, behavior and person.

Early on I stressed the "Safety Triad," with behavior and person sides representing the psychology of safety. That's the focus of this book. But don't overlook the need for environmental change. The environment includes physical conditions and the general atmosphere or ambiance regarding safety. For example, safety can be considered a top-down "condition of employment" or a bottom-up "opportunity to get involved." Which of these perspectives you choose influences your safety-related attitudes and behaviors.

Environment, behavior, and person factors are dynamic and interactive; a change in one eventually impacts the other two. For example, behaviors that reduce the probability of injury often involve environmental change, and lead to attitudes consistent with the safe behaviors. In other words, when people choose to act safely, they act themselves into safe thinking and this often results in some environmental change.

PRINCIPLE 8: Don't count on common sense for safety improvement.

Most common sense is not common. It's biased by our subjective interpretation of unique experiences. As a researcher of psychological principles for more than 35 years, I have become quite committed to this basic principle (discussed in Chapter 2). Indeed, I've dedicated most of my career to discovering principles of human behavior through systematic application of the scientific method. You've probably noticed by now that I get quite disturbed when I read or hear "pop psychology" based on unfounded intuition or "common sense." Many statements I've read or heard related to the psychology of safety sound good but are incorrect. Profound knowledge comes from rigorous research and theory development, and often runs counter to common sense.

PRINCIPLE 9: Safety incentive programs should focus on the process rather than outcomes.

One of the most frequent common-sense mistakes in safety management is the use of outcome-based incentive programs. Giving rewards for avoiding an injury seems reasonable and logical. But it readily leads to covering up minor injuries and a distorted picture of safety performance. The basic activator-behavior-consequence contingency (see Chapter 8 and Principle 19) demonstrates that safety incentives need to focus on process activities, or safety-related behaviors.

PRINCIPLE 10: Safety should not be considered a priority, but a value with no compromise.

As discussed in Chapter 3, this is the ultimate vision: Safety becomes a value linked to every priority in the workplace, or wherever we find ourselves. Priorities change according to circumstances; values are deep-seated personal beliefs beyond compromise. Establishing safety as a value won't happen overnight. It will take time, and there is no "quick fix."

PRINCIPLE 11: Safety is a continuous fight with human nature.

I know people who meet the behavioral criteria for holding safety as a value—they practice safety, teach it, go out of their way to actively care for the safety of others. But their numbers are few. Why? Because human nature (or natural motivating consequences, Chapter 10) typically encourages at-risk behavior. The soon, certain, positive, and natural consequences of risky behavior are hard to overcome. We're talking about comfort, convenience, and expediency. Now consider safe alternatives, which often mean

Figure 17.1. Safe behavior often requires us to resist natural motivation for comfort, convenience, and efficiency.

discomfort, inconvenience, and inefficiency. The inconveniences involved in safely locking out equipment are illustrated in Figure 17.1. When you compete with natural supportive consequences in order to teach, motivate, or change behavior, you're fighting human nature.

Principle 12: Behavior is learned from three basic procedures: classical conditioning, operant conditioning, and observational learning.

Through naturally occurring consequences and planned instructional activities, we learn every day, and we develop attitudes and emotional reactions to people, events, and environmental stimuli. The mechanisms for learning voluntary and involuntary behavior and emotions were reviewed in Chapter 8.

The critical aspect of this principle is that our actions and feelings result from what we learn through experience, both planned and unplanned. Basic learning principles can be applied to change what we do and feel. But experience and practice develop habits that are hard to break. Plus, it's possible that natural contingencies and social influences support a "bad habit" or "negative attitude." So, learning new behaviors and attitudes often requires another fight with human nature.

PRINCIPLE 13: People view behavior as correct and appropriate to the degree they see others doing it.

Since personal experience often convinces us that "it's not going to happen to me," we need a powerful reason to perform safely when personal injury is improbable. So consider this: Everyone who sees you acting safely or at risk either learns a new behavior or thinks what you're doing is OK. Now consider the vast number of people who observe your behavior every day. Our influence as a social model gives us special responsibility to go out of our way for safety.

PRINCIPLE 14: People will blindly follow authority, even when the mandate runs counter to good judgment and social responsibility.

This principle was discussed in Chapter 5 as a potential barrier to safe work practices. The fact that people often follow top-down rules without regard to potential risk is alarming. This puts special responsibility on managers and supervisors who give daily direction. These front-line leaders could signal, even subtly, the approval of at-risk behavior in order to reach production demands. And people are apt to follow even implicit demands from their supervisor to whom they readily delegate responsibility for injury that could result from at-risk behavior.

PRINCIPLE 15: Social loafing can be prevented by increasing personal responsibility, individual accountability, group cohesion, and interdependence.

This principle was introduced in Chapter 14 when discussing ways to increase group productivity and synergy. Giving up personal responsibility for safety to another person (Principle 14) could be due to the social mechanisms presumed to influence social loafing. It's possible to decrease blind compliance to rules that foster at-risk behavior by manipulating factors found to decrease social loafing. Thus, workplace interventions and action plans need to be implemented with the aim of increasing people's perceptions of personal responsibility and accountability, as well as their sense of group cohesion and interdependence.

PRINCIPLE 16: On-the-job observation and interpersonal feedback is key to achieving a Total Safety Culture.

Critical behavior checklists (Chapter 8) and communicating the results of checklist observations (Chapter 11) put this principle to work. The penguin in Figure 17.2 does not have sufficient equipment nor a method to fly. "Try harder" won't work without the right process. Interpersonal observation and feedback are basic tools to use when the safety director says, "Try harder."

Figure 17.2. Best efforts fail without the right equipment and method.

PRINCIPLE 17: Behavior-based safety is a continuous DO IT process.

D = Define target behaviors, *O* = Observe target behaviors, *I* = Intervene to improve behaviors, and *T* = Test impact of intervention. Motivation to "try harder" cannot work without appropriate method. As Dr. Deming (1991) put it, "Goals without method—what could be worse?" This four-step process enables continuous improvement through an objective behavior-focused approach. As detailed in Chapter 8, people need to decide on critical target behaviors to observe. After baseline observations are taken, an intervention is developed and implemented. By continuing to observe the target behaviors, the impact of the intervention program can be objectively evaluated. Results might suggest a need to refine the intervention, carry out another one, or define another set of behaviors to work on. The next four principles provide guidance for designing behavior-change interventions.

PRINCIPLE 18: Behavior is directed by activators and motivated by consequences.

As explained in Chapters 8, 9 and 10, external or internal events (referred to as activators) taking place before behavior only motivate to the extent that they signal or specify consequences. Intentions and goals can motivate behavior if they stipulate positive or negative consequences. Understanding this principle is critical to developing effective behavior-change techniques. Chapter 9 showed how this principle guides the development of more effective activators, and Chapter 10 outlined procedures for improving the motivational power of consequences.

PRINCIPLE 19: Intervention impact is influenced by: amount of response information, participation and social support, and external consequences.

Interventions that give specific instructions (response information) and get participants actively involved are likely to influence behavior and attitude change. And, if the intervention facilitates support from others, such as co-workers or family members, it can have lasting effects. Furthermore, people are more apt to develop internal motivation when external rewards or threats are relatively small and insufficient to completely justify the target behavior. This phenomenon has been referred to as the "less-leads-to-more effect" and is most likely to occur when people feel personally responsible for their choice of action and the resulting consequences.

PRINCIPLE 20: Extra and external consequences should not overjustify the target behavior.

The various examples of positive consequences presented in Chapters 10 and 11 (from thank-you cards to the privileges, commendations, and small tangibles) are not large nor expensive. For reasons discussed above, rewards should not provide complete justification for desired behavior. We don't want people complying with safety rules only to gain a reward or avoid a penalty. If that's the case, what happens when we take away the consequence, good or bad? We take away the reason to comply. This is why many people wear PPE at work, but rarely at home.

PRINCIPLE 21: People are motivated to maximize positive consequences (rewards) and minimize negative consequences (costs).

This principle offers another reason why people are not likely to follow safe operating procedures in the absence of external controls, or behavior-consequence contingencies. As reflected in Principle 11, natural external consequences usually support risk-taking at the expense of safe alternatives, which are usually more inconvenient, uncomfortable, or time consuming.

Of course, this principle relates to many behaviors. In Chapter 12, it was used to explain why people often don't rush to help in a crisis. If there are more perceived costs than benefits to intervening, actively caring behavior is unlikely. Therefore, a prime strategy for increasing safety and actively caring behaviors is to overcome the costs (negative consequences) with benefits (positive consequences). Various kinds of consequences are defined by the next principle.

PRINCIPLE 22: Behavior is motivated by six types of consequences: positive vs. negative, natural vs. extra, and internal vs. external.

Understanding these characteristics (as detailed in Chapter 10) can enable significant insights into the motivation behind observed behavior. Appreciating these various consequences can also suggest whether external intervention is called for to change behavior, and what kind of intervention to implement. It's possible the natural external consequences supporting ongoing at-risk behavior cannot be overcome with extra external consequences. In this case, long-term behavior change requires the modification of the natural consequences or the application of techniques discussed in Chapter 14 to alter internal consequences. Throughout this text, I have downplayed the use of negative consequences, and the reasons are reflected in the next principle.

PRINCIPLE 23: Negative consequences have four undesirable side effects: escape, aggression, apathy, and counter-control.

How did you feel the last time you received a reprimand from a supervisor? Maybe you felt like slinking away, or taking a swipe at him. Chances are you didn't go back to the job charged up. Perhaps you wanted to do something to make him look bad. These and other undesirable side effects of using negative consequences are discussed in Chapter 10.

PRINCIPLE 24: Natural variation in behavior can lead to a belief that negative consequences have more impact than positive consequences.

As detailed in Chapter 11, behavior fluctuates from good to bad for many reasons. Peak performance seldom can be sustained, and poor performance is almost bound to get better at some point. So if you praise someone and their performance falters, don't swear off positive feedback. And don't overestimate the power of your reprimand if it gets some immediate results. Keep things in perspective.

PRINCIPLE 25: Long-term behavior change requires people to change "inside" as well as "outside."

The psychology of safety requires us to consider both external behavior and internal person factors. Chapter 13 focused on the role of person states in influencing people to actively care for another person's safety and health. Chapter 14 showed how outside factors can be manipulated to influence these person states, and thus increase actively caring behavior. A Total Safety Culture requires integrating both behavior-based and person-based psychology. The next several principles focus on understanding "inside" factors.

PRINCIPLE 26: All perception is biased and reflects personal history, prejudices, motives, and expectations.

Appreciating this principle is key to understanding people, and realizing the importance of actively listening to others before intervening. It also supports the need to depend on objective, systematic observation for knowledge rather than common sense (Principle 8).

It's important to realize the reciprocal relationship between perception and behavior. Perceptions influence actions, and in turn actions influence perceptions. If we perceive risk, we'll act to reduce it; by acting to reduce risk, we'll become more aware of other risks.

The current popularity of safety perception surveys in industry reflects an increased awareness of how perceptions impact safety performance. These surveys can help pinpoint issues that need attention, and activities in need of an intervention. They can also be used to assess the person factors influenced by a particular intervention program.

PRINCIPLE 27: Perceived risk is lowered when a hazard is perceived as familiar, understood, controllable, and preventable.

When people perceive a new risk, they adjust their behavior to avoid it. Call it "fear of the unknown." The reverse is also true. As discussed in Chapter 5, research has shown that hazards perceived as familiar, understood, controllable, and preventable are viewed as less risky. This is why many hazards are underestimated by employees.

PRINCIPLE 28: The slogan "all injuries are preventable" is false and reduces perceived risk.

Frankly, I believe that telling people all injuries are preventable insults their intelligence. They know better. It's difficult enough to anticipate and control all environmental and behavioral factors contributing to injuries, but controlling factors inside people is clearly impossible.

Such a slogan can make it embarrassing to report an injury, and could influence a cover-up. "If they think all injuries are preventable, they will think I was really stupid to have this injury, so I'd better not report it." The most critical problem with this popular slogan is that it can reduce the perception of risk. Hazards considered controllable and preventable are perceived as relatively risk free.

PRINCIPLE 29: People compensate for increases in perceived safety by taking more risks.

As reviewed in Chapter 5, researchers have shown that some people will compensate for a decrease in perceived risk by performing more risky behavior. In other words, some people increase their tolerance for risk when feeling protected with a safety device. As shown in Figure 17.3, high technology safety engineering can give a false sense of security. This is not the case for people who hold safety as a value (Principle 10).

GEEZ! RELAX, WILL YA … IF THE COMPUTERS SENSORS SAY THERE'S NO OBSTRUCTION AHEAD, THERE'S NO OBSTRUCTION AHEAD.

Figure 17.3. Sometimes high technology causes reduced perception of risk and at-risk behavior.

PRINCIPLE 30: When people evaluate others they focus on internal factors; when evaluating personal performance, they focus on external factors.

As discussed in Chapter 6, this principle is termed "The Fundamental Attribution Error." It contributes to systematic bias whenever we attempt to evaluate others, from completing performance appraisals to conducting an injury investigation. Because we are quick to attribute internal (person-based) factors to other people's behavior, we tend to presume consistency in others because of permanent traits or personality characteristics. To explain injuries to other persons we use expressions like, "He's just careless," "She had the wrong attitude," and "They were not thinking like a team."

On the other hand, when evaluating our own behavior, we point the finger to external factors. This should make us stop and realize the many observable variables that can be measured and often changed to increase everyone's safety-related behavior and reduce injuries throughout a culture.

PRINCIPLE 31: When succeeding, people over-attribute internal factors; but when failing, people over-attribute external factors.

This research-based principle is referred to as the "self-serving bias" (see Chapter 6), and is sure to warp injury investigations. Placing blame for a mistake on outside variables is just a basic defense to protect one's self-esteem. In most organizations, even a minor injury is perceived as a failure. As a result, the victim is sure to avoid discussing inside, person factors contributing to the mishap. Statements like "I was fatigued," "I didn't know the proper procedure," or "My mind was on other things" are far less probable than "The work demands were too severe," "The trainer didn't show me the correct procedure," or "Excessive noise and heat distracted me." My advice is to accept the self-serving bias and allow people their ego-protecting excuses. Then search for measurable external factors (including behaviors) that can be changed to reduce the probability of another injury.

PRINCIPLE 32: People feel more personal control when working to achieve success than when working to avoid failure.

The sense of having control over life events is one of the most important person states contributing to our successes and failures. When we feel in control we're more motivated and work harder to succeed. We're also more likely to accept failure as something we can change. Thus, the value of increasing people's sense of personal control over safety is obvious. One way to do this is to develop scoring procedures for safety achievements, rather than focusing on the number of reported injuries as a measure of success. This puts the emphasis on measuring process activities that can lead to loss control or injury prevention, as detailed in Chapter 15.

PRINCIPLE 33: Stressors lead to positive stress or negative distress, depending on appraisal of personal control.

When we believe we can do things to reduce our stressors—work demands, interpersonal conflict, boredom—we're more motivated to take control. As discussed in Chapter 6, this is positive stress, an internal person state not nearly as detrimental to safety as distress. We feel distress when we believe there is little we can do about current stressors. This state can lead to frustration, exhaustion, burnout, and dangerous behavior.

It's important to recognize states of distress in others and attempt to help them. After actively listening to another person's concerns, you might be able to offer constructive suggestions. Sometimes it's useful to help people distinguish between the stressors they can control and the ones they can't. We can be concerned about a lot of things, but we can only control some of these.

Helping people focus on the stressors they *can* reduce builds their sense of self-efficacy, personal control, and optimism. These are the person states that imply empowerment and increase one's willingness to actively care: "I can make a difference."

PRINCIPLE 34: In a Total Safety Culture everyone goes beyond the call of duty for the safety of themselves and others—they actively care.

Here we have a primary theme of this book. While behavior-based psychology provides methods and techniques to improve the human dynamics of safety, principles from person-based psychology need to be considered to ensure that behavior-based tools are used. The ultimate aim is to integrate behavior-based and person-based psychology so everyone participates in efforts to achieve a Total Safety Culture. In the ideal culture, everyone actively cares for the safety and health of others.

PRINCIPLE 35: Actively caring should be planned and purposive, and focus on environment, person, or behavior.

Part Four of this book is all about actively caring for safety, from understanding why people resist it (Chapter 12) to implementing strategies that increase it (Chapter 14). We need to plan ways to enable and nurture as much actively caring behavior as possible, rather than sit back and wait for "random acts of kindness."

By considering the three domains of actively caring focus, we can sometimes get more benefit from an act of kindness. In particular, including behavior-focused actively caring can often result in the most benefit. For example, it's often possible to include specific behavioral advice, direction, or motivation with a donation (environment-focused actively caring) and with crisis intervention and active listening (person-focused actively caring).

PRINCIPLE 36: Direct, behavior-focused actively caring is proactive and most challenging, and requires effective communication skills.

Some acts of caring are relatively painless and effortless—contributing to a charity, sending a get-well card, or actively listening to another person's problems. But telling someone how to change their behavior can be confrontational and challenging, especially when it's direct. Think of the parent telling the child: "I want to discuss your behavior." This is the type of active caring we're most likely to avoid, which is unfortunate because it's the most beneficial. But even parents will pass up chances to talk about behavior with their kids. Why?

You often hear parents lament that they had no training in how to raise children. Our resistance is partly due to lack of confidence in our communication skills. Proper training and practice as a safety coach increases our ability to actively care for safety in this most beneficial way.

PRINCIPLE 37: Safety coaching that starts with Caring and involves Observing, Analyzing, and Communicating, leads to Helping.

The basic components of effective safety coaching were presented in Chapter 11, with the letters of COACH signifying labels for the sequential events in the process. The coaching process should start with an atmosphere of interpersonal Caring and an agreement that the coach can Observe an individual's performance, preferably with a behavioral checklist. Then the coach Analyzes the observations from a fact-finding, system-level perspective. Subsequently, the results are Communicated in one-to-one actively caring conversation, with the sole purpose to Help another individual reduce the possibility of personal injury.

PRINCIPLE 38: Actively caring can be increased indirectly with procedures that enhance self-esteem, belongingness, and empowerment.

This principle reflects one of the most innovative and critical theories discussed in this book (Chapter 13). Substantial research is available to support each component of this principle. Procedures that enhance a person's sense of self-esteem ("I am valuable"), belongingness ("I belong to a team"), and empowerment ("I can make a difference") make it more likely that person will actively care for the safety or health of another person. Nourishing each of these person states leads to the actively caring belief that "We can make valuable differences."

PRINCIPLE 39: Empowerment is facilitated with increases in self-efficacy, personal control, and optimism.

This principle was mentioned earlier when reviewing the distinction between stress and distress (Principle 33). When people's sense of self-efficacy ("I can do it"), personal control ("I am in control"), or optimism ("I expect the best") is increased, they are less apt to experience distress and more likely to feel empowered ("I can make a difference"). In addition, empowerment increases one's inclination to perform actively caring behaviors.

Note that empowerment does not necessarily result from receiving more responsibility. In order to truly feel more empowered, people need to perceive they have the skills, resources, and opportunity to take on the added responsibility (self-efficacy), believe they have personal impact over their new duties (personal control), and expect the best from their efforts to be more responsible (optimism).

PRINCIPLE 40: When people feel empowered their safe behavior spreads to other situations and behaviors.

In a Total Safety Culture people go beyond the call of duty for safety. This means they perform safe behaviors in various situations. More specifically, they show both stimulus generalization—performing various particular safe behavior in various settings—and response generalization—performing safe behaviors related to a particular target behavior. Both types of generalization occur naturally when safety becomes a value rather than a priority (Principle 10). Obviously we need to intervene in special ways to promote safety as a value. Our applied research has shown that facilitating empowerment is one special way to increase generalization and cultivate safety as a value.

PRINCIPLE 41: Actively caring can be increased directly by educating people about factors contributing to bystander apathy.

In Chapter 14, I discussed strategies for encouraging actively caring behavior directly. This principle expresses the most basic procedure for doing this. Research has shown that educating people about the barriers to helpful behavior can remove some obstacles and increase the probability of actively caring behavior. Similarly, I have found that discussing the barriers to safe behavior can motivate people to improve safety, provided they also learn specific techniques for doing this.

PRINCIPLE 42: As the number of observers of a crisis increases, the probability of helping decreases.

This principle, supported with substantial behavioral research, is probably the first barrier to actively caring behavior that should be taught. It's strange but true, and means that people cannot assume that someone else will intervene in a crisis. In fact, the most common excuse for not acting is probably something like, "I thought someone else would do it" or "I didn't know it was my responsibility." This principle reflects the need to promote a norm that it's everyone's responsibility to actively care for safety. We can never assume someone else will correct an at-risk behavior or condition.

PRINCIPLE 43: Actively caring behavior is facilitated when appreciated and inhibited when unappreciated.

Making an effort to actively care directly for someone else's safety is a big step for many people and deserves genuine recognition. Then, if advice is called for to make the actively caring behavior more effective, corrective feedback should be given appropriately. Be sure to make your deposits first. All actively caring behavior is well-intentioned, but not frequently practiced with the kind of feedback that shapes improvement. A negative reaction to an act of caring can be quite punishing, and severely discourage a person from trying again. Consequently, much of the future of actively caring behavior is in the hands of those who receive people's attempts to actively care.

PRINCIPLE 44: A positive reaction to actively caring can increase self-esteem, empowerment, and belongingness.

This is a follow-up to Principle 43, and supports the need to sincerely recognize occurrences of actively caring behavior. Although research in this area is lacking, it is intuitive that feeling successful at actively caring behavior should lead to more active caring. Success should enhance self-esteem, empowerment, and belongingness, and so indirectly increase the probability of more caring acts. Thus, we have the potential for a mutually supporting cycle of actively caring influence, provided the reactions to actively caring behavior are positive.

PRINCIPLE 45: The universal norms of consistency and reciprocity motivate everyday behaviors, including actively caring.

These social influence norms have powerful impact on human behavior. Sometimes people apply these norms intentionally to influence others. At other times, these norms are activated without our awareness. Regardless of intention or awareness, behavior-change techniques derived from these norms can be very effective. In Chapter 14, I discussed how these social influences can be used to directly increase actively caring behavior. The next three principles reflect that information.

PRINCIPLE 46: Once people make a commitment, they encounter internal and external pressures to think and act consistently with their position.

This is why I say you can act people into thinking differently, or think people into acting differently. If people act in a certain way on the "outside," they will adjust their "inside"—including perceptions, beliefs, and attitudes—to be consistent with their behaviors. The reverse is also true, but throughout this text I've recommended targeting behavior first because it's easier to change on a large scale. As presented in Chapters 9 and 10, we know much more about changing behavior than perceptions, beliefs, and attitudes, because behavior is easier to measure objectively and reliably.

PRINCIPLE 47: The consistency norm is responsible for the impact of "foot-in-the-door" and "throwing a curve."

As detailed in Chapter 14, the "foot-in-the-door" and "curve ball" techniques of social influence succeed because of the consistency norm (Principle 46). When an individual agrees with a relatively small request, for example, to serve on a safety committee, you have your foot in the door. To be consistent, the person is more likely to agree later with a larger request, perhaps to give a safety presentation at a plant-wide meeting. Similarly, when people sign a petition or promise card that commits them to act in a certain way, say to actively care for the safety of others, they experience pressure from the consistency norm to follow through.

The technique of "throwing a curve" occurs when a person is persuaded to make a particular decision because there's not much at stake. Then the stakes are raised. Due to the consistency norm, the individual will likely stick with the original decision. Here's a safety application: An employee is asked to serve on a safety committee. No big deal—the committee meets just once a month. But then the employee is asked to attend more meetings because a special project has come up. To remain true to his first decision, the employee will probably stay on the committee and take on the additional work.

PRINCIPLE 48: The reciprocity norm is responsible for the impact of the door-in-the-face technique.

The reciprocity norm is a powerful determinant of human behavior. Its influence is reflected in the popular expression, "one good turn deserves another" and in the well-known Golden Rule. This is another reason to actively care for safety. One good act will likely lead to another.

The success of the "door-in-the-face" technique depends on the reciprocity norm. If an employee shuts the door on a major request, he's more likely to be open to a lesser request. If you ask for something less imposing, costly, or inconvenient after the initial refusal, your chances of being accepted are greater than if you started with the minor request. Your willingness to withdraw the larger request sets up an obligation to reciprocate and accept the smaller request. In Chapter 14, I discussed applications of this principle to promote actively caring behavior.

PRINCIPLE 49: Numbers from program evaluations should be meaningful to all participants and direct and motivate intervention improvement.

The last two principles relate to the critical issue of program evaluation (Chapter 15). In safety, the total recordable injury rate (TRIR) is the most popular evaluation number used to rank companies for safety rewards. It is calculated by multiplying the number of workplace injuries by 200,000 and dividing the answer by the total person-hours worked in that time period. What an obvious example of an abstract number with little meaning. The most direct measure of ongoing safety performance comes from behavioral observations, and in Chapters 8, 11, and 15, I recommended ways to obtain meaningful feedback numbers from such process evaluation.

In Chapters 13 and 15, I presented questionnaires that assess particular person states to gauge reactions to interventions. Such evaluation tools are not as objective and directly applicable to process improvement as feedback charts from behavioral observations. But results of surveys to measure perceptions, attitudes, or person states can be meaningful to program participants if explained properly. If given before and during an intervention process, questionnaires can reflect changes in the "inside" factors that impact program acceptance, participation, and future success.

PRINCIPLE 50: Statistical analysis often adds confusion and misunderstanding to evaluation results, thereby reducing social validity.

Complex statistics are appropriate and often necessary for research journals. But if the purpose of a program evaluation is to improve a safety process, we need to provide numbers that give the most meaningful feedback to program participants—the people in the best position to improve the process.

Recall also the lesson from Drs. Sandman (1991) and Slovic (1991) that group statistics have minimal impact on risk perception (Chapter 5). If your goal is to increase risk awareness and motivate safe behavior, the most influential evaluation tool you can use is actually anecdotal. The most stimulating feedback usually comes from the personal report of an injured employee. However, as illustrated in Figure 17.4, the victim might want to cover up at-risk behavior leading to an injury. The culture needs to support reporting personal injuries, as well as discussing ways to prevent future incidents.

As illustrated in Figure 17.5, when people give personal testimony, the presentation is more influential than a statistical analysis. We should proba-

Figure 17.4. At-risk behavior is embarrassing if it results in an injury.

Figure 17.5. Personal testimony is a more powerful motivator than group statistics.

bly spend less time calculating summary injury statistics and more time eliminating the barriers to the personal reporting and investigating of safety-related incidents—from near hits and first-aid cases to lost-time injuries.

In Conclusion

This chapter reviews the principles of human dynamics discussed throughout this book. Founded on research published in scientific journals, they enable profound understanding of the psychology of safety. Use them as guidelines to develop, implement, evaluate, and refine safety-improvement programs and you'll make a positive difference in the safety of your organization, community, or culture.

Champions are needed to lead this process. Some are easy to find; others will evolve when the principles reviewed here are taught. Give potential champions opportunities to teach these principles and help develop interventions. Active participation increases both belief in the principles and empowerment to apply them to achieve a Total Safety Culture.

There is no quick-fix to culture change. The journey is not to be without bumpy roads, forced detours, and missed turns. These principles are your map to reach an enviable destination, but be prepared to blaze new paths and traverse difficult terrain. And please don't forget to take a break now and then to appreciate journey milestones. Recognize behaviors that contribute to a successful journey.

At the end of the second Deming workshop I attended, a participant raised his hand to ask one final question. When acknowledged, he stood and walked to the nearest microphone and stated, "Dr. Deming, you have taught us many important principles to consider when designing procedures to transform a culture. But frankly, the challenge seems overwhelming. Can we really expect to make a difference in our lifetime?"

W. Edwards Deming, at age 92, replied, "That's all you've got!"

References

American Heritage Dictionary (1991). Second College Edition. New York: Houghton-Mifflin Company.

Asch, S. E. (1956). Studies of independence and conformity: A minority of one against a unanimous majority. *Psychological Monographs, 70,* 1–70.

_____. (1958). Effects of group pressure upon modification and distortion of judgments. In E. E. Maccoby, T. M. Newcomb, & E. L. Hartley (Eds.). *Readings in social psychology* (Third Edition). New York: Holt, Rinehart & Winston.

Broadbent, D., Cooper, P. F., Fitzgerald, P., & Parker, K. (1982). The cognitive failures questionnaire (CFQ) and its correlates. *British Journal of Clinical Psychology, 21,* 1–16.

Brown, H. J., Jr., (1991). *Life's little instruction book.* Nashville, TN: Rutledge Hill Press.

Bugelski, B. R., & Alimpay, D. A. (1961). The role of frequency in developing perceptual sets. *Canadian Journal of Psychology, 15,* 205–211.

Cialdini, R. B. (1993). *Influence: Science and practice,* Third Edition. New York: Harper Collins College Publishers.

Covey, S. R. (1989). *The seven habits of highly effective people: Restoring the character ethic.* New York: Simon and Schuster.

_____. (1990). *Principle-centered leadership.* New York: Simon and Schuster.

Daniels, A. C. (1994). *Bringing out the best in people.* New York: McGraw-Hill, Inc.

Deming, W. E. (1985). Transformation of western style of management. *Interfaces,* 15(3), 6–11.

_____. (1986, July). *Drastic changes for western management.* Abstract for the meeting of TIMS/ORS at Gold Coast City, Australia.

_____. (1986). *Out of crisis.* Cambridge, MA: Massachusetts Institute of Technology, Center for Advanced Engineering Study.

_____. (1991, May). *Quality, productivity, and competitive position.* Four–day workshop presented in Cincinnati, Ohio, by Quality Enhancement Seminars, Inc.

_____. (1992, January). *Instituting Dr. Deming's methods for management of productivity and quality.* Two–day workshop presented in Washington, D. C. by Quality Enhancement Seminars, Inc.

_____. (1993). *The new economics for industry, government, education.* Cambridge, MA: Massachusetts Institute of Technology, Center for Advanced Engineering Study.

Editors of Conari Press (1993). *Random acts of kindness.* Emeryville, CA: Conari Press.

Frankl, V. (1962). *Man's search for meaning: An introduction to logotherapy.* Boston: Beacon Press.

Fulghum, R. (1988). *All I really need to know I learned in kindergarten.* New York: Ivy Books.

Geller, E. S. (1994). Ten principles for achieving a Total Safety Culture. *Professional Safety,* 39(9), 18–24.

Geller, E. S. (1996). *The psychology of safety: How to improve behaviors and attitudes on the job.* Radnor, PA: Chilton Book Company.

Geller, E. S., Roberts, D. S., & Gilmore, M. R. (1996). Predicting propensity to actively care for occupational safety. *Journal of Safety Research, 27,* 1–8.

Guastello, S. J. (1993). Do we really know how well our occupational accident prevention programs work? *Safety Science, 16,* 445–463.

Haddon, W., Jr. (1968). The changing approach to the epidemiology, prevention, and amelioration of trauma: The transition to approaches etiologically rather than descriptively based. *American Journal of Public Health, 58,* 1431–1438.

Hall, E. T. (1966). *The hidden dimension.* Garden City, NY: Doubleday.

Hansen, L. (1994). Rate your B.O.S.S.—Benchmarking organizational safety strategy. *Professional Safety, 39*(6), 37–43.

Hayakawa, S. I. (1978). *Language in thought and action* (Fourth Edition). New York: Harcourt Brace Jovanovich, Publishers.

Heinrich, H. W., Peterson, D., & Roos, N. (1980). *Industrial accident prevention: A safety management approach* (Fifth Edition). New York: McGraw-Hill, Inc.

Latané, B., & Darley, J. M. (1968). Group inhibition of bystander intervention. *Journal of Personality and Social Psychology, 10,* 215–221.

_____ & _____. (1970). *The unresponsible bystander: Why doesn't he help?* New York: Appleton-Century-Crofts.

Maier, S. F., & Seligman, M. E. P. (1976). Learned helplessness: Theory and evidence. *Journal of Experimental Psychology:* General, 105, 3–46.

Maslow, A. H. (1943). A theory of human motivation. *Psychological Review, 50,* 370–396.

_____. (1971). *The farther reaches of human nature.* New York: Viking.

Milgram, S. (1963). Behavioral studies of obedience. *Journal of Abnormal and Social Psychology, 67,* 371–378.

_____. (1974). *Obedience to authority.* New York: Harper Collins.

The New Merriam-Webster Dictionary (1989). Springfield, MA: Merriam-Webster, Inc., Publishers.

Pavlov, I. P. (1927). *Conditional reflexes.* (G. V. Anrep, Ed. and translator). London: Oxford University Press.

Peck, M. S. (1979). *The different drum: Community making and peace.* New York: Simon and Schuster.

Piliavin, J. A., Dovidio, J. F., Gaertner, S. L., & Clark, R. D., III. (1981). *Emergency intervention.* New York: Academic Press.

Rotter, J. B. (1996). Generalized expectancies for internal versus external control of reinforcement. *Psychological Monographs, 80,* No. 1.

Sandman, P. M. (1991). *Risk = Hazard + Outrage: A formula for effective risk communication.* Videotaped presentation for the American Industrial Hygiene Association. Environmental Communication Research Program, P.O. Box 231, Cook College, Rutgers University, New Brunswick, NJ.

Seligman, M. E. P. (1975). *Helplessness: On depression development and death.* San Francisco, CA: Freeman.

Selye, H. (1976). *Stress without distress.* Philadelphia, PA: Lippincott.

Skinner, B. F. (1938). *The behavior of organisms.* Acton, MA: Copley Publishing Group.

_____. (1953). *Science and human behavior.* New York: Free Press.

_____. (1974). *About behaviorism.* New York: Alfred A. Knopf.

_____. (1981). Selection by consequences. *Science, 213,* 502–504.

Slovic, P. (1991). Beyond numbers: A broader perspective on risk perception and risk communication. In D. G. Mayo & R. D. Hollander (Eds.), *Deceptable evidence: Science and values in risk management*. New York: Oxford University Press.

Steel, S. (1996) (Ed.). *Evaluating the effectiveness of agricultural safety and health initiatives*. Itasca, IL: The National Safety Council.

Storey, C. (1989). Excellence: It's in the cards. *Training & Development Journal*, September, 46–48.

Toronto Sun (1977, July 26th). Reports from insurance/accident forms. Toronto, Ontario, Canada.

Wilde, G. J. S. (1994). *Target risk*. Toronto, Ontario, Canada: PDE Publications.

Yerkes, R. M., & Dodson, J. D. (1908). The relation of strength of stimulus to rapidity of habit formation. *Journal of Comparative and Neurological Psychology*, 18, 459–482.

Index